T0304514

Predictive Hydrology

A Frequency Analysis Approach

Predictive Hydrology

A Frequency Analysis Approach

Paul Meylan
Member of Board of Directors
AIC Ingénieurs Conseil S.A.
Lausanne, Switzerland

Anne-Catherine Favre
École Nationale Supérieure de l'Énergie
l'Eau et l'Environnement (ENSE3)/Institut National
Polytechnique de Grenoble (GINP)
France

André Musy
École Polytechnique Fédérale de Lausanne (EPFL)
Switzerland

CRC Press
Taylor & Francis Group
an Informa business
www.taylorandfrancisgroup.com

6000 Broken Sound Parkway, NW
Suite 300, Boca Raton, FL 33487
711 Third Avenue
New York, NY 10017
2 Park Square, Milton Park
Abingdon, Oxon OX14 4RN, UK

Science Publishers
Jersey, British Isles
Enfield, New Hampshire

Published by Science Publishers, an imprint of Edenbridge Ltd.

- St. Helier, Jersey, British Channel Islands
- P.O. Box 699, Enfield, NH 03748, USA

E-mail: *info@scipub.net* Website: *www.scipub.net*

Marketed and distributed by:

ISBN: 978-1-57808-747-1

Cover illustrations: Swiss Air Force: river *Klein Melchaa,* Village of Giswil (OB) August 2005 flood, overtopping with fluvial spreading

Library of Congress Cataloging-in-Publication Data

Meylan, Paul.
Predictive hydrology:a frequency analysis approach/Paul Meylan, Anne-Catherine Favre, Andre Musy.
p. cm.
Includes bibliographical references and index.
ISBN 978-1-57808-747-1 (hardcover : alk. paper)
1. Hydrometeorology. 2. Hydrology--Statistical methods. 3. Frequency curves. 4. Hydrology--Mathematical models. 5. Hydrological forecasting. I. Favre, Anne-Catherine. II. Musy, A. III. Title.
GB2803.2.M48 2012
551.57--dc23

2011038912

Published by arrangement with Presses polytechniques et universitaires romandes, Lausanne, Switzerland

Translation of: *Hydrologie Fréquentielle, une science prédictive,* Presses polytechniques et universitaires romandes, Lausanne, Switzerland, 2008.

French edition: © Presses polytechniques et universitaires romandes, Lausanne, 2008. ISBN: 978-2-88074-797-8

Printed in the United States of America

Foreword

Statistical hydrology has long played a critical role in civil and environmental engineering practice and in the design of water resource structures. Hydrology has also played a role in the development of statistics, particularly in extremes. For example, one of the earliest applications of the statistics of extreme values was in the definition and practical computations of the return period of floods. An enormous amount of information can be obtained from the statistical analysis of time series of hydrological and meteorological measurements whose diverging rescaled range with sample size has been pointed out -- notably, by a hydrologist -- as the signature of how Nature works.

The fundamental questions addressed in this book concern how to assign a probability to the future occurrence of an event of a given magnitude. Engineering practice involves designing and operating hydraulic structures for water resource exploitation, which have as inputs random events, not deterministic phenomena, as is the rule in other areas of engineering. A wide range of periods of drought and flooding occur across the Earth's surface. And hydrological extreme events are expected to increase due to the acceleration of the hydrologic cycle as a result of global warming. The role of frequency analysis and the study of trends in hydrology (study of non-stationarity) remain ever more important, with substantial social implications, chiefly associated with current questions of sustainability. The authors of this fine book have clearly laid out the principles of frequency analysis in hydrology at a level suitable for majors in civil and environmental engineering that will beautifully integrate with their other courses in probability and statistics, hydrology, and water resources management.

Topics thoroughly treated include probabilistic definitions, risk and data analysis, model selection and parameter estimation, and uncertainty studies, as well as a forward-looking introduction to Bayesian methods, non-stationarity and the use of copulas in hydrology. These topics are discussed in the last chapter and provide an original launching point for further studies in statistical hydrology.

This book will serve as an important point of reference and should be part of the library of practicing engineers and scientists. The authors have combined their many years of experience in practice, teaching and research in statistical hydrology to produce this superb textbook that will be appreciated by both students and teachers.

Professor Marc Parlange
School of Architecture, Civil and
Environmental Engineering, EPFL

Preface

Data about the natural environment of any region are a vital part of its heritage and provide the most accurate assessment of its condition. Therefore this information is absolutely essential for the engineers, managers and politicians involved in any development or protection efforts. Acquiring this data is an exacting task, and requires significant organization and investment at every step in the process. Because such information concerns systems or processes that are extremely complex, it is both very valuable, and quite rare. In general and no matter the amount, this data – usually local data – is insufficient to provide a reliable representativeness of the system from which it originates and of its space-time evolution. This means that great caution must be used in analyzing the data in order to extract all the information, sometimes hidden, that they contain.

Frequency analysis serves this purpose. It is one of the methods and techniques of statistical analysis that are essential before data can be used for specific applications. More precisely, frequency analysis makes possible the projection of data in probabilistic space in order to foresee the behavior and reactions of the systems or processes the data describes or represents, so that decision-makers will have credible and substantiated information before taking any action. For example, before beginning work on any flood management structure, dike, impounding reservoir, or water diversion, it is essential to have relevant and solid information for assessing possible extraordinary rainfall events that could impact the watershed concerned. The appropriate frequency analysis makes it possible to specify the criteria necessary for designing such hydraulic structures, and also provides an estimation of the quality of the information available by determining its degree of certainty or likelihood.

These methods and techniques, often resulting from experience, are based on statistical principles and approaches. The intuition, experience and scientific knowledge of the engineer must be confirmed, expanded and justified using a rigorous approach. After this, however, the results must be tested and checked experimentally. This is because in actual practice, some of the statistical tests that are recommended by the theorists often reveal themselves to be poorly suited for checking certain characteristics of time series, auto-correlation in particular.

This book combines experimental approaches and those from scientific developments, which makes it somewhat unique. The methods and approaches contained herein appeared first in teaching materials in the form of course notes. Originally, the material was based on case studies, which was reworked for teaching purposes and designed for students in applied engineering, the water sciences in particular. Paul Meylan, my unfailing friend on this project, made a significant contribution in developing this instructional material and converting it into the appropriate scientific language. Afterwards, Anne-Catherine Favre, a mathematician and statistician, brought some careful and authoritative improvements to the analysis and its rigor and introduced some very promising fields of research. The end result is a book that combines the very different training and backgrounds of the authors, and especially their distinctive scientific and teaching interests. However, this book would never have been completed without the enormous dedication and perseverance of the two scientists named above, who conducted research under my supervision during several years at the Laboratoire Hydrologie et Aménagements HYDRAM (now ECOL : Ecological Engineering Laboratory) at the École Polytechnique Fédérale de Lausanne. For this I will be forever grateful.

This book is slanted towards the hydrological sciences and their applications in terms of water and environmental engineering. It describes the treatment of all sorts of data, but focuses especially on data about meteoric and terrestrial water, with an emphasis on rare and unusual events, or what are commonly referred to as "extreme events". This book is aimed particularly at students in civil engineering, environmental sciences and technology, hydrology, geography, geology and ecology, all of whom must be concerned with the judicious use of data for the design and construction of water management structures. This book will also be useful for engineering practitioners, who are faced constantly with problems regarding data, or the lack thereof, when attempting to develop an accurate portrayal of hydrological processes and to analyze their behavior. And finally, this book will be useful for teachers in disciplines involving the processing of environmental data or any other data concerning the evolution over time of a system (data linked to traffic, air quality, water quality, etc. ...).

The topics covered in this book are by no means exhaustive, but they do provide some answers to the main questions that arise during the design or restoration of water management structures or when assessing their impacts on the natural or built environment. This book also addresses some difficult problems, such as the non-stationarity of natural phenomena or the multidimensional aspects of risk, and provides some alternative solutions or at least suggests some pathways for finding a resolution. My hope is that readers will find the methods and techniques they need for the successful realization of projects related to water management and connected fields, and thus contribute to the sustainable development of our societies and our environment.

André Musy
Honorary Professor, EPFL

TRANSLATION

Given the success of the French-language versions of this book, which was first published in 2008, we received a great deal of encouragement to translate it into English for the benefit of a broader readership. This we were able to do with the help of translation and revision by Ms Robyn Bryant and especially the committed involvement of the co-authors. Financial support for this translation was provided by the Institut d'ingénierie de l'environnement (Institute of Environmental Engineering) at EPFL and The Federal office of environment of Switzerland, Hydrological division at Bern. The authors are deeply grateful to these individuals and institutions and to the administrative structures where these various individuals are involved, namely the AIC consulting firm in Lausanne.

Contents

CHAPTER 1

Introduction

1.1 FORECASTING AND PREDICTION

Two very different approaches are commonly used in hydrology concerning the future – either the *forecast*, which is generally applied for relatively short periods of time, or the *prediction*, for the longer term. This distinction corresponds to two different approaches, but also to different problems (Morel-Seytoux, 1979).

The first approach could be applied to the *operation* of hydraulic structures or systems, while the second concerns the *planning* of installations and the design of structures.

1.1.1 Forecasting

Weather forecasting is probably the most popular example of the concept of forecasting: a study of synoptic situations makes it possible to establish forecasts for the following day or days.

Similarly, for the purpose of managing water projects, observed conditions (current or past) are projected into the relatively near future, for example by applying a hydraulic routing model.

In this kind of problem, the question being asked is along the lines of: *"What is the discharge that will flow through this particular hydraulic structure tomorrow at noon?"* The forecasting approach plays a key role not only in the management of water resources (such as irrigation) but also in providing warnings about floods and other natural disasters like inundations.

1.1.2 Prediction

The concept of prediction can be illustrated by the familiar example of someone planning their next ski holiday. They know from experience (which is to say, through observation of past situations) that they are more likely to encounter the appropriate snow conditions for their favorite sport during February, rather than December.

A similar approach is used for the purposes of planning and designing hydraulic structures: the future evolution of the process under study is described based on an analysis of past measurements, but only in terms of *probability of occurrence.*

The question to be answered in this kind of problem is along the lines of: "*What is the probability that a discharge of 1340 [m³/s] will be reached or exceeded in the next fifteen years?*" The precise date of the event being considered is not a concern in this approach.

1.2 FREQUENCY ANALYSIS

Frequency analysis is a statistical method of prediction that consists of studying past events that are characteristic of a particular hydrological (or other) process in order to determine the probabilities of occurrence of these events in the future.

This prediction is based on defining and implementing a *frequency model,* which is a mathematical description of the statistical behavior of a random variable through its probability distribution function $F(x) = \Pr(X \le x)$. Such models describe the probability of occurrence of an event equal to or less than a given value. Figure 1.1 illustrates graphically such a model. For example, one could determine that the probability of observing a daily rainfall greater than x [mm] is p%. The Gumbel distribution is the most common frequency model used in hydrology. It is expressed as:

$$F(x) = \exp\left[-\exp\left\{-\frac{(x-\alpha)}{\beta}\right\}\right] \tag{1.1}$$

In this model, $F(x)$ is the probability distribution function, also known as distribution function or more commonly, *cumulative frequency.* α and β are the two parameters of the model: α is the location parameter and β is the scale parameter.

The probability of observing an annual event greater than the value x is in this case $p(x) = 1 - F(x)$: this statistical quantity is called the probability of exceedance or exceedance probability.

In hydrology we often use the concept of return period $T(x) = 1/p(x)$ rather than that of cumulative frequency. Return period, which is a function of $p(x)$, will be defined in greater detail in Chapter 2.

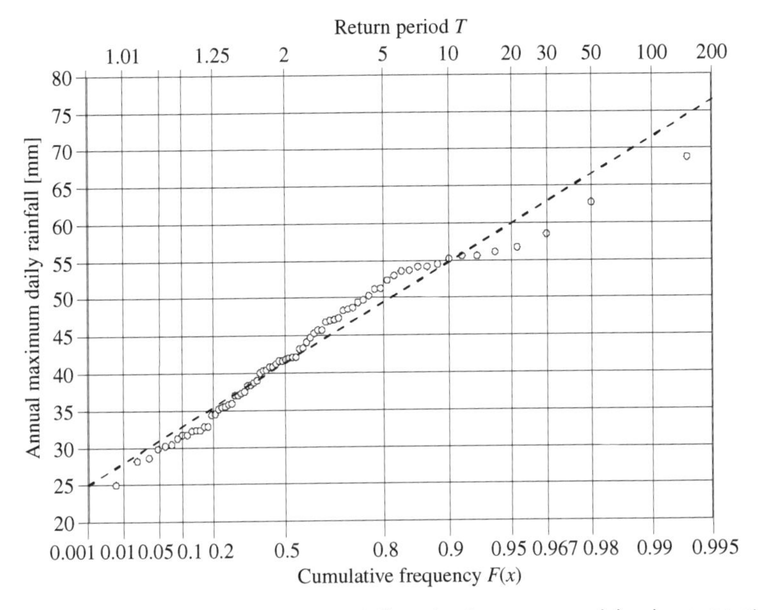

Fig. 1.1 Example of a graphical representation of a frequency model: rain-gauge station at Bex (Switzerland) 1930-2004.

The calibration of the model $F(x)$ serves to estimate the value of the model's parameters, for example α and β in equation (1.1). Then, it is possible to use the model to determine, for example, the quantile x corresponding to a given return period.

Frequency analysis implies a whole range of processing methods and statistical techniques and constitutes a complex system that this book aims to explain. The various steps in frequency analysis can be summarized graphically in a very simple way, as shown in Figure 1.2.

Although frequency analysis can be applied to any kind of data, we will concentrate on the problem that hydrologists most often face, which is the estimation of the probability of rare events (often referred to as *extreme* events).

1.3 THE PRINCIPLE OF FREQUENCY ANALYSIS

This section presents in a very simplified way the required approach to carry out a frequency analysis corresponding in this case to the methodology used to build a *frequency model*. We will look first at two simple and empirical approaches, and in later chapters we will turn our attention to a more detailed study of the various steps (see Fig. 1.2).

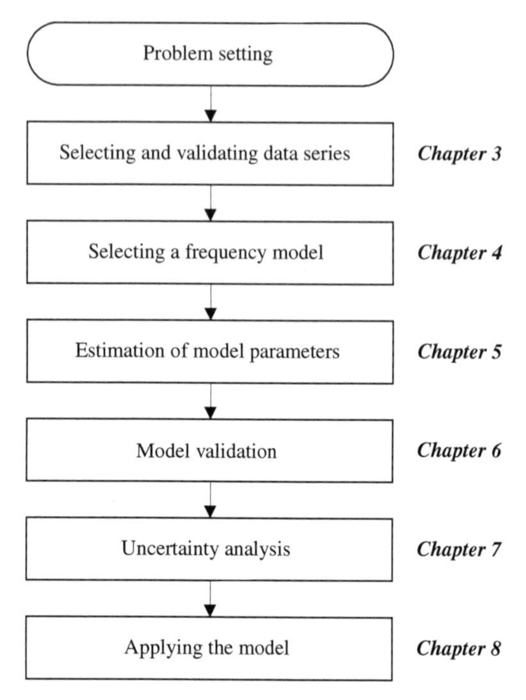

Fig. 1.2 The main steps in a frequency analysis.

1.3.1 A Simplified Modeling Approach: The Empirical Frequencies Polygon

For the purposes of illustration, let us consider the mean daily discharges denoted by Q. Table 1.1 shows the annual maximum mean daily discharges for the Massa River at Blatten near Naters (Switzerland) from 1922 to 2003. Note that the years 1929 and 1930 are missing.

Table 1.1 Annual maximum mean daily discharge of the Massa at Blatten (1922-2003) in [m^3/s].

Date	$Q_{max\ yearly}$	Date	$Q_{max\ yearly}$	Date	$Q_{max\ yearly}$
07-09-1922	78	07-31-1951	63	08-08-1978	51
08-16-1923	108	08-16-1952	73	07-31-1979	67
07-17-1924	78	08-12-1953	62	08-04-1980	64
07-23-1925	69	08-07-1954	66	08-17-1981	66
08-23-1926	72	07-19-1955	58	08-17-1982	72
08-07-1927	75	08-10-1956	60	07-21-1983	84
07-17-1928	95	07-06-1957	59	07-13-1984	73
07-01-1931	61	08-02-1958	73	08-23-1985	57

Table 1.1 Contd. ...

Table 1.1 Contd. ...

08-22-1932	70	07-11-1959	65	08-05-1986	68
08-13-1933	81	08-28-1960	57	08-18-1987	83
07-12-1934	83	08-11-1961	67	07-27-1988	72
07-01-1935	84	08-03-1962	65	07-24-1989	79
07-27-1936	70	07-23-1963	67	08-03-1990	73
07-26-1937	66	07-19-1964	73	08-12-1991	76
08-21-1938	71	06-28-1965	68	08-21-1992	94
08-06-1939	72	08-14-1966	57	07-06-1993	67
07-27-1940	65	08-03-1967	73	08-08-1994	91
07-26-1941	68	07-11-1968	71	07-23-1995	73
08-30-1942	74	08-15-1969	73	07-24-1996	73
08-19-1943	76	08-08-1970	70	08-06-1997	55
09-01-1944	97	08-20-1971	85	08-23-1998	92
07-22-1945	87	08-15-1972	82	08-07-1999	85
08-06-1946	79	08-22-1973	68	10-15-2000	88
08-04-1947	95	08-17-1974	63	08-04-2001	98
08-09-1948	65	08-07-1975	61	06-24-2002	88
08-10-1949	80	07-17-1976	67	08-04-2003	98
07-02-1950	74	07-13-1977	51		

The first step in our analysis consists of studying the *distribution* of this sample, which could be done for example by drawing a *histogram,* which is the most usual graphical representation of the distribution.

The number of intervals k to consider can be determined by applying Sturges' rule (Sturges, 1926) or Iman-Conover's (Iman and Conover, 1983). For the sample in Table 1.1, the data is split into 6 intervals as shown in Table 1.2.

Table 1.2 Splitting of annual maximum mean daily discharges [m^3/s] in intervals of the same amplitude.

Class	Intervals	Counts n_j	Cumulated counts N_j	Cumulative frequency F_j
1	[50-60]	8	8	8/80 = 10.00%
2	[60-70]	25	33	33/80 = 41.25%
3	[70-80]	27	60	60/80 = 75.00%
4	[80-90]	11	71	71/80 = 88.75%
5	[90-100]	8	79	79/80 = 98.75%
6	[100-110]	1	80	80/80 = 100.00%

The resulting histogram aims to display the shape of the distribution of values and especially to assess the degree of symmetry. The histogram in Figure 1.3, for example, reveals a positive asymmetry. The data are skewed to the right.

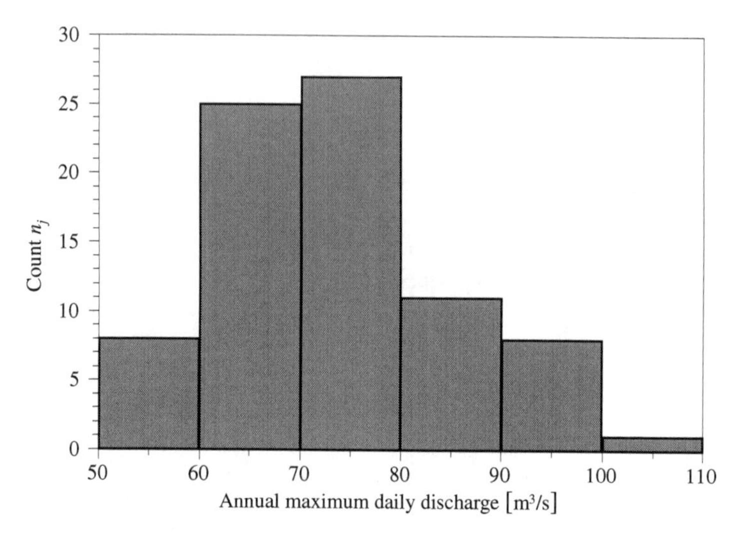

Fig. 1.3 Histogram of annual maximum mean daily discharge [m^3/s].

It is worth noting that the choice of the boundary points for each interval, which is highly arbitrary has a great influence on the final shape of the histogram, and this effect is more evident when the sample size is small.

When the sample shows a high asymmetry, it is useful to split the sample into intervals with unequal widths. It is important to remember that it is the *density of counts for each class*[1] that has to be plotted in the *y*-axis.

The histogram provides a first rough approximation of the frequency of occurrence of a particular event. In the case of the Massa at Blatten, we can see that the annual maximum mean daily discharge is larger than 90 [m^3/s] nine times out of 80, and is between 70 and 80 [m^3/s] 27 times out of the total of 80.

A more convenient way to estimate frequencies is to draw a cumulative frequency polygon. This is created using a splitting of the data as in Table 1.2, and by considering the cumulative counts N_j of each class. These quantities can be converted easily to cumulative frequencies by computing the ratio N_j/N where N is the sample size. The last column in Table 1.2 shows the cumulative frequency of the selected

[1] The *density of counts* n'_j in an interval is the ratio of the counts n_j in interval j to the width (or amplitude) of this same interval. The basic rule applied to histograms is that the *area* of each of the rectangles is proportional to the number of counts.

intervals. A graphical representation (a cumulative frequency polygon) is drawn by plotting the upper limit of each interval on the x-axis and the corresponding cumulative frequency on the y-axis. A point with a null cumulative frequency ($F = 0$) corresponding to the lower limit of the first non-empty interval completes the diagram.

From the cumulative frequency polygon (Fig. 1.4) it is easy to deduce the frequency of non-exceedance corresponding to a given value of the variable under study: for example, in our illustrating case, the annual maximum mean daily discharge of the Massa at Blatten did not exceed 80 [m^3/s] 75 times out of 100.

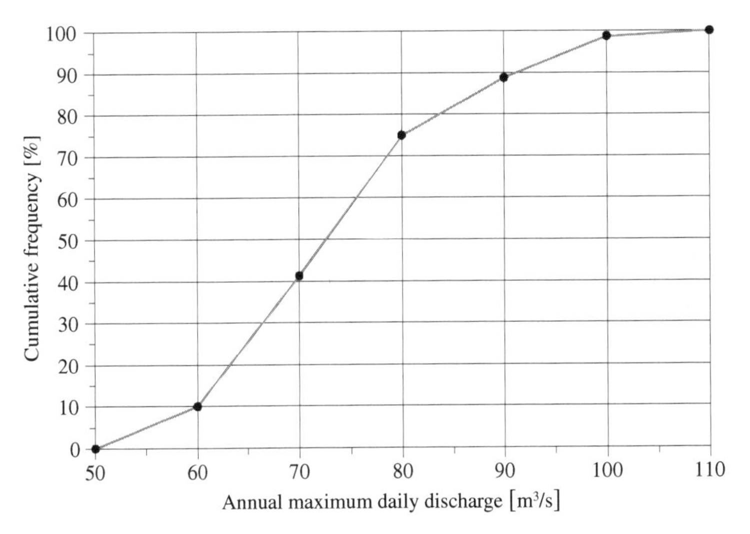

Fig. 1.4 Cumulative frequency polygon of the annual maximum mean daily discharges [m^3/s].

The *frequency of exceedance,* which is the complement probability of the cumulative frequency, usually labeled as p, is defined by:

$$p(x) = 1 - F(x) \tag{1.2}$$

This means than the annual maximum mean daily discharge of the Massa at Blatten has exceeded 100 [m^3/s] only 1.25 times out of 100.

The drawing of a cumulative frequency polygon, as described above, is a common technique of descriptive statistics.

1.3.2 A Simplified Modeling Approach: Calculating Empirical Frequency

Hydrologists, who often have to deal with small sample sizes, still prefer to use a *"comprehensive" form* of this statistical technique. The latter relies on choosing intervals that contain only one individual, so that *all the information is used and retained.* Consequently and in general these intervals are of variable width but have the same counts. Consequently they have the same probability of occurrence (the reason why they are called *equiprobable*).

The one and only difficulty in this kind of situation involves adequately defining the boundary points of the different intervals.

The construction of a (comprehensive) frequency polygon is based on a preliminary classification of the values in the sample in ascending (or descending) order. The order number of a value in a sample ranked in ascending order corresponds to the rank denoted by r. The corresponding value, or the value of rank order r, is denoted $x_{[r]}$. Table 1.3 shows the sorting by *ascending values* of the annual maximum mean daily flows of the Massa at Blatten.

Table 1.3 Sample of sorted annual maximum mean daily discharges [m^3/s].

Rank	Year	$Q_{max\ yearly}$	Rank	Year	$Q_{max\ yearly}$
1	1978	51	...		
2	1977	51	71	2002	88
3	1997	55	72	1994	91
4	1985	57	73	1998	92
5	1960	57	74	1992	94
6	1966	57	75	1928	95
7	1955	58	76	1947	95
8	1957	59	77	1944	97
9	1956	60	78	2001	98
10	1975	61	79	2003	98
11	1931	61	80	1923	108

The next step consists of defining the cumulative empirical frequency $\hat{F}$ associated with each value of the sample x. Here, we will discuss only four possible approaches. The difference between these approaches lies in the choice of the bounds for the equiprobable intervals.

Approach 1

This method consists of putting each of the n values at the upper bound of each of the n equiprobable intervals (Fig. 1.5).

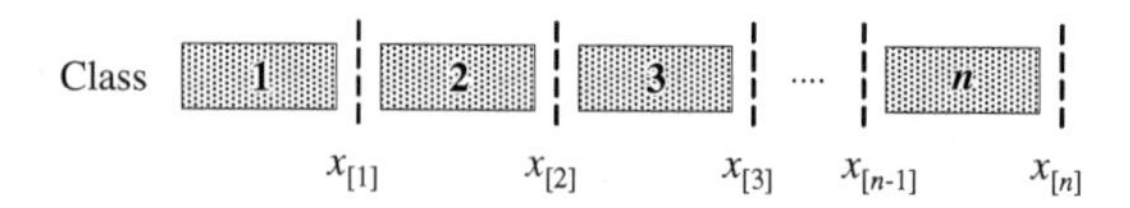

Fig. 1.5 First approach for defining intervals.

This leads to the following formula:

$$\hat{F}(x_{[r]}) = \frac{r}{n} \tag{1.3}$$

which provides an empirical cumulative frequency in the ascending sorted value $x_{[r]}$, and is consistent with the "classical" approach in statistics.

Approach 2

If the sorted values $x_{[r]}$ are no longer considered as upper bounds but as lower bounds of each equiprobable interval (Fig. 1.6), the equation for estimating the cumulative frequency becomes:

$$\hat{F}(x_{[r]}) = \frac{r-1}{n} \tag{1.4}$$

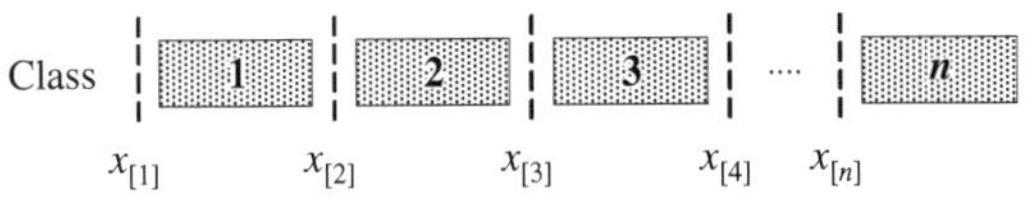

Fig. 1.6 Second approach for defining intervals.

Table 1.4 compares the empirical cumulative frequencies obtained by applying these two approaches. We can see that the values obtained by applying equations (1.3) and (1.4) are not *symmetric* around the median value (the average of the values corresponding respectively to rank 40 and 41 in our illustrating example, as the sample size is 80).

Table 1.4 Comparison of empirical cumulative frequencies obtained with equations (1.3) and (1.4).

Rank *r*	*r/n*	*(r–1)/n*
1	0.0125	0
2	0.025	0.0125
3	0.0375	0.025
...		
40	0.5	0.4875
41	0.5125	0.5
...		
78	0.975	0.9625
79	0.9875	0.975
80	1	0.9875

This asymmetry can be avoided by changing the way of determining the bounds of the various intervals. This leads to the two following approaches.

Approach 3

This method is based on the rule applied in the first approach but with the addition of a fictitious interval in the upper part of the distribution, in order to obtain a symmetric empirical cumulative frequency (Fig. 1.7).

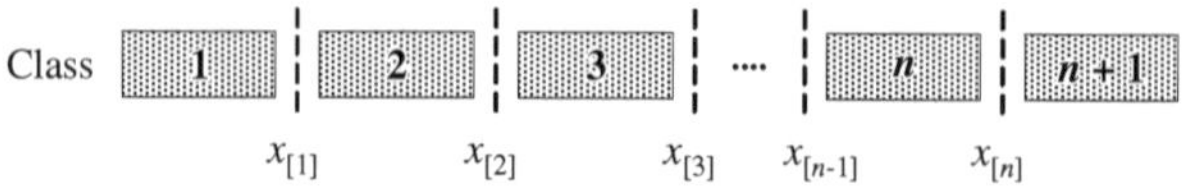

Fig. 1.7 Third approach for defining equiprobable intervals.

This method leads to the following formula, known as the *Weibull formula*:

$$\hat{F}(x_{[r]}) = \frac{r}{n+1} \tag{1.5}$$

Approach 4

The last possibility discussed here consists of choosing an intermediate formula between the first two strategies, which means putting each of the x values building the sample in the middle of each equiprobable interval[2]:

$$\hat{F}(x_{[r]}) = \frac{r-0.5}{n} \tag{1.6}$$

This relation (equation (1.6)) is known as the Hazen formula.

Table 1.5 shows the cumulative frequencies obtained using the last two formulas. As can be seen, they result in symmetry around the median.

Table 1.5 Comparison of empirical cumulative frequencies obtained with equations (1.5) and (1.6).

Rank r	$r/(n+1)$ (Weibull)	$(r-0.5)/n$ (Hazen)
1	0.012	0.006
2	0.025	0.019
3	0.037	0.031
...		
40	0.494	0.494
41	0.506	0.506
...		
78	0.963	0.969
79	0.975	0.981
80	0.988	0.994

[2] Purists should forgive this shortcut.

In frequency analysis applied to hydrology which we call "frequential hydrology," this type of formula is often referred to as a *plotting position formula*. A pragmatic solution to the question of choosing the most appropriate empirical cumulative frequencies among the various equations exposed in the hydrological literature will be outlined in Chapter 5. This chapter deals with the parameters estimation technique, which is based on empirical frequency.

Figure 1.8 shows a cumulative frequency polygon of the annual maximum mean daily discharges of the Massa at Blatten obtained by applying the Hazen formula (Eq. (1.6)).

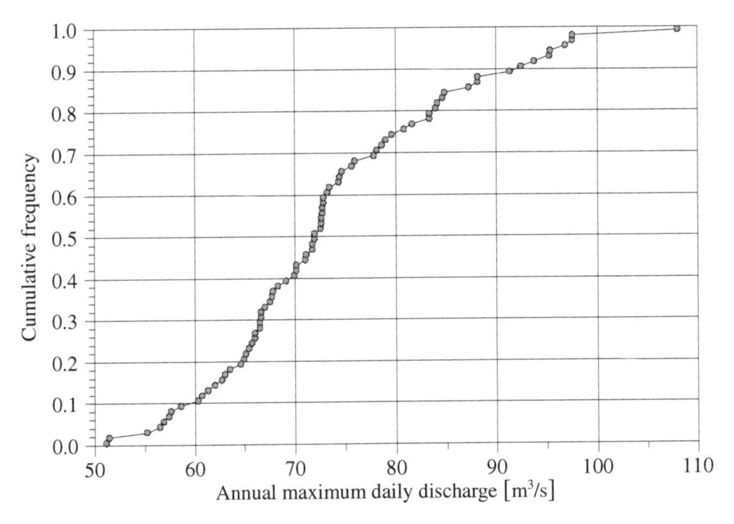

Fig. 1.8 Empirical cumulative frequency distribution of the annual maximum mean daily discharges [m^3/s] (obtained with the Hazen formula).

Although the empirical method explained above allows us to answer a number of questions, it is usually more interesting to apply a *frequency model* (in other words, a probability distribution function) and to estimate its parameters from the sample of interest in order to be able to apply the model, and in particular to extrapolate quantiles beyond the observations.

1.3.3 Model Calibration

Questions concerning the *calibration*[3] of a model (in other words, parameters estimation) have given rise to numerous papers in the hydrological literature. This subject will be discussed further in

[3] also use the terms *specification*

Chapter 5. For now, we will expose without additional comments the fitted model for the sample of the annual maximum mean daily discharges of the Massa at Blatten from 1922 to 2003 (Table 1.1), which is a Gumbel distribution with parameters $\hat{\alpha} = 68.71$ and $\hat{\beta} = 9.16$ (see Eq. (1.1)):

$$\hat{F}(x) = \exp\left[-\exp\left\{-\frac{(x-68.71)}{9.16}\right\}\right] \quad (1.7)$$

Contrary to the common practice in statistics, with *hydrological frequency diagrams* the usual practice is to plot the cumulative frequency on the x-axis and the variable under study on the y-axis (Fig. 1.9).

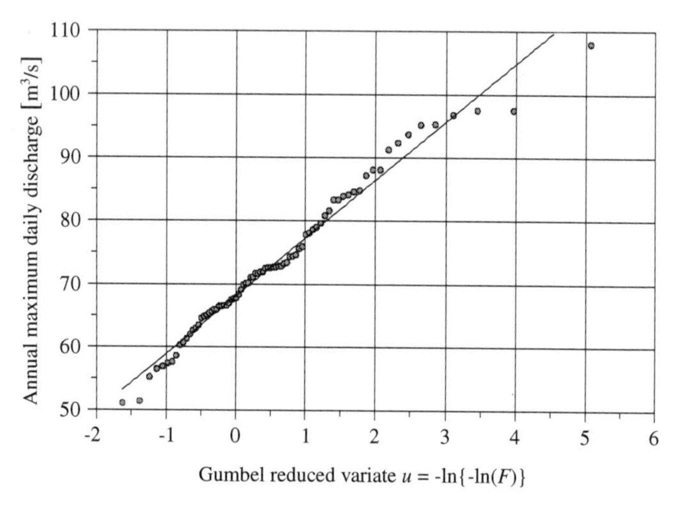

Fig. 1.9 Gumbel distribution fitted to the sample of annual maximum mean daily discharges of the Massa at Blatten.

1.3.4 Applying a Frequency Model

Once a model has been defined and validated (see the next chapters), it can be *applied* in order to compute the quantile corresponding to a given return period or, conversely, to compute the return period of specified events[4]. It can also be used to compare the behavior of several samples, or as data input for stochastic models.

In our example, as the variables of interest are the *annual* maxima, equation (1.7) makes it possible to compute the annual maximum mean daily discharge corresponding to a given cumulative frequency $\hat{F}(x)$. This quantile is obtained simply by inverting equation (1.7):

---

[4] The reader is referred to Chapter 2 for the explanation of the return period concept

$$\hat{x}_q = 68.17 + 9.16\left[-\ln\left\{-\ln(\hat{F})\right\}\right] \quad (1.8)$$

Then it is easy to estimate the quantile of the discharge corresponding to a given cumulative frequency, which is, as we shall explain in the following chapter, directly related to the return period of the considered event.

1.4 OUTLINE OF THIS BOOK

This book is divided into three main sections:

The first section consists of a very general introduction (the current chapter) aiming to frame the topic, followed by a more in-depth discussion in Chapter 2 of the concept of *return period* and its links with the *probability* of events. In Chapter 2 we will also introduce the concept of risk and attempt to clarify its components.

In the second section we will expose the basics of the subject by explaining the different steps in a *frequency analysis* as was shown in Figure 1.2. (Chapters 3 to 8).

The third section (Chap. 9) puts into perspective some of the open questions along with recent and ongoing developments.

CHAPTER 2

Return Period, Probability and Risk

In this chapter we show that the probability of occurrence of an event being greater than a given magnitude should never be the sole criterion on which to base a decision to execute a project or upgrade an existing project, and should never be the sole criterion for designing a structure. Once we have exposed this issue, we look at the concept of *risk* in relatively summary fashion, but with enough detail to understand the fundamental aspects of this very important idea, which is complementary to the concept of probability. We will also clarify as precisely as possible some of the terms and concepts connected to the notion of risk, because many of these related words are often poorly understood, or interpreted differently in various references. This is also why, for reasons of clarity, we use the term *probability of failure* in the second half of this chapter, and keep the term *risk* for later on when we have defined it more precisely in Section 2.3.

But before we get to this concept of risk and its related context, it is helpful to define the concept of *return period*, as we have mentioned in the previous chapter. We also explore how return period is computed using probabilities.

2.1 RETURN PERIOD

The concept of return period is well known in the hydrological community. Hydrologists are also frequently misunderstood by other scientists simply because they fail to clearly explain that return period is not related to calendar time but is a probabilistic concept. This confusion is also induced by the fact that the term *period of occurrence* is sometimes used. We are emphasizing this aspect so that, at least, hydrologists do not make this same error themselves!

2.1.1 Event

An occurrence is described as an *event of interest* when a random variable, characteristic of a hydrological process, is equal to or greater than an arbitrary threshold x_T:

$$\{X > x_T\}$$

We should mention here that the magnitude of a rare event (corresponding to a very high threshold) is inversely proportional to its probability of occurrence: events of very large magnitude take place less frequently than more moderate events. Frequency analysis aims to link a hydrological event with its frequency of occurrence by means of probability distributions.

2.1.2 Recurrence Interval

The recurrence interval Y is the time separating two occurrences of an event $\{X > x_T\}$.

Example

Let us consider again the annual maximum mean daily discharges of the Massa River at Blatten (Switzerland) from 1922 to 2003 (see Table 1.1). We fix a threshold of x_T = 90 [m^3/s] and we want to calculate the average recurrence interval.

Figure 2.1 illustrates the data for the Massa River at Blatten as a time series plot.

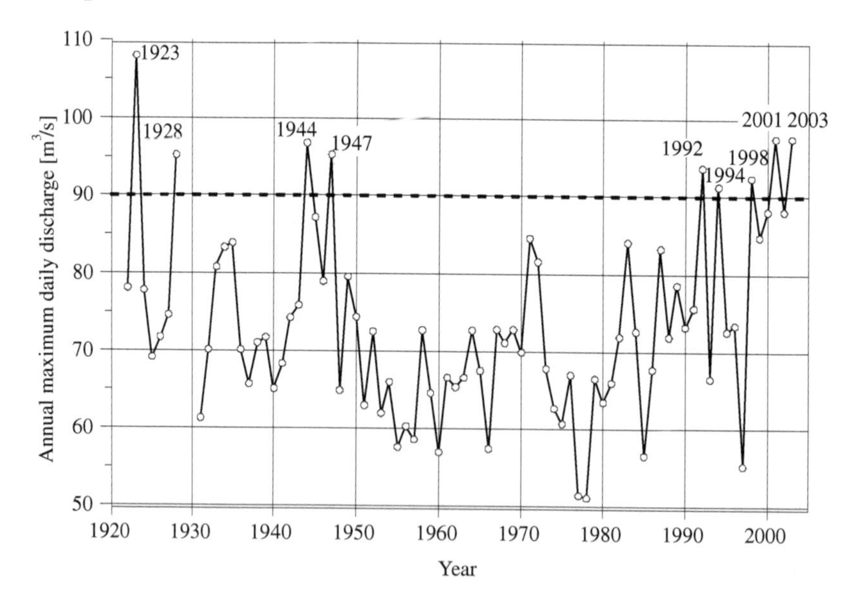

Fig. 2.1 Time series plot of the annual maximum daily discharges of the Massa River at Blatten. The threshold of x_T = 90 [m^3/s] is indicated with the dotted line.

Table 2.1 summarizes the recurrence intervals for this sample.

Table 2.1 Recurrence intervals of events $\{X > x_T\}$ in the case of the annual maximum mean daily discharges of the Massa River at Blatten. The selected threshold here is x_T = 90 [m^3/s]. The years 1929 and 1930 are missing and we are assuming that for these two years $X \leq 90$.

1922		1923		1928		1944		1947		1992		1994		1998		2001		2003
	1		5		16		3		45		2		4		3		2	

We have $y \in [1, 45]$ and

$$\bar{y} = \frac{1 + 5 + 16 + 3 + 45 + 2 + 4 + 3 + 2}{9} = \frac{81}{9} = 9$$

In our example, the average recurrence interval is 9 years.

2.1.3 Definition of Return Period

The return period T of the event $\{X > x_T\}$ is the expectation of the random variable Y representing the recurrence interval measured for a great number of occurrences:

$$T(x_T) = \mathrm{E}[Y] \tag{2.1}$$

In terms of probabilities only two cases are possible for the event of interest:

- The event $\{X > x_T\}$ occurs; we call this a "success" and its probability is equal to p;
- The complement of this event is $\{X \leq x_T\}$; we call this a "failure" and its probability is 1-p.

Since we have only two possible cases here, the random variable corresponding to the event is distributed according to a *Bernoulli distribution.* To compute the probability that the interval of recurrence is equal to y, or $\Pr(Y = y)$, we only have to remember that we have (y-1) *consecutive* "failures" followed by a "success." As the trials are independent, the probability can be written as:

$$\underbrace{(p-1)(p-1)\ldots(p-1)}_{y-1}\, p = (1-p)^{y-1}\, p \tag{2.2}$$

Equation (2.2) corresponds to the frequency function of a random geometric variable of parameter p which models *the number of trials (tests) required to reach the occurrence of the event knowing that the probability of success in a single trial is p.*

To determine return period, then by definition, it is sufficient to compute its expectations as follows:

$$
\begin{aligned}
\mathrm{E}[Y] &= \sum_{y=1}^{\infty} y(1-p)^{y-1}\, p \\
&= p + 2(1-p)\,p + 3(1-p)^2\,p + 4(1-p)^3\,p + \ldots \\
&= p\{1 + 2(1-p) + 3(1-p)^2 + 4(1-p)^3 + \ldots\} \\
&= p\left\{\sum_{i=0}^{\infty} -(1-p)^{i+1}\right\}' \qquad (2.3) \\
&= p\left(\frac{p-1}{1-(1-p)}\right)' \\
&= \frac{p}{p^2} \\
&= \frac{1}{p}
\end{aligned}
$$

As can be shown, simple algebraic computations lead to an infinite sum corresponding to a geometric series with a common ratio (1-p).

Thus we obtain the following formal definition:

$$T(x_T) = \frac{1}{\Pr(X > x_T)} = \frac{1}{1 - \Pr(X \le x_T)} = \frac{1}{p(x_T)} = \frac{1}{1 - F(x_T)} \qquad (2.4)$$

The return period $T(x_T)$ is thus defined as the inverse of the probability of exceedance $p(x_T) = 1 - F(x_T)$ where $F(x_T)$ is the cumulative frequency. The rarer the event is, the lower its probability of exceedance and therefore the greater its return period. Considering a very long period of time, the event with a value superior to x_T will occur *on average* once every T years.

It should be pointed out that by definition, as the return period is the expectation of the recurrence interval, it possesses a unit of measurement (year, trimester, month or any other measurement of duration). In hydrology, the reference period is usually a year, corresponding to a hydrological cycle. When infra-annual series are introduced in order to increase the quantity of information (see Chapter 8), the distribution has to be corrected to bring it back to an "annual" period.

To conclude this definition, it is worth mentioning again the fact that the return period is the expectation of the recurrence interval Y of events $\{X > x_T\}$. Consequently, the concept of return period does not imply any regularity in time or involve any cyclical notion of the incidence of events.

2.2 THE PROBABILITY OF EVENTS

We have shown in the preceding paragraphs that the event of interest $\{X > x_T\}$ can be modeled with a Bernoulli random variable. There are two other discrete statistical distributions associated with Bernoulli's: the binomial distribution and the geometric distribution.

The binomial distribution models the number of successes or failures resulting from N Bernoulli trials.

The geometric distribution is suitable when the variable of interest is the number of Bernoulli trials requested to obtain the first success. We see below that these two distributions are involved in computing given probabilities linked to return period.

First of all, let us suppose, for example, that we are interested in a quantile of flood corresponding to a 50 years return period:

$$T = 50,\ \Pr(X > x_{50}) = p(x_{50}) = \frac{1}{50} = 0.02,\ F(x_{50}) = 1 - 0.02 = 0.98$$

2.2.1 Probability that at least One Value Exceeds x_T in N Successive Years

If we denote, for example, the number of discharges exceeding the critical value x_T in 30 years as W, the resulting random variable follows a binomial distribution $\beta(30, 0.02)$. Then:

$$\begin{aligned}
\Pr(W \geq 1) &= 1 - \Pr(W = 0) \\
&= 1 - \binom{30}{0} 0.02^0 (1 - 0.02)^{30} \\
&= 1 - 0.98^{30} \\
&= 0.45
\end{aligned}$$

In a general case we have $W \sim \beta(N, p)$. Therefore the probability is computed as follows:

$$\begin{aligned}
\Pr(W \geq 1) &= 1 - \Pr(W = 0) \\
&= 1 - \binom{N}{0} p^0 (1 - p)^N \\
&= 1 - (1 - p)^N
\end{aligned}$$

We can also use the following general equations:

$$\Pr(W \geq 1) = 1 - (1 - p)^N$$
$$\Pr(W \geq 1) = 1 - \left(1 - \frac{1}{T}\right)^N \tag{2.5}$$

A Helpful Consequence

It is therefore possible to compute the probability, denoted here as p_1, of observing at least one flood with a return period of T years during a period of T years by the relation $p_1 = 1 - \left(1 - \frac{1}{T}\right)^T$. Table 2.2 shows the resulting probabilities for several return periods ranging from 2 to 1000 years.

Table 2.2 Probability of observing at least one flood with a return period of *T* years during a period of *T* years.

T	2	5	10	50	100	1000
p_1	0.750	0.672	0.651	0.636	0.634	0.632

For $T = 2$ years, the probability is high ($p_1 = 0.75$). To reduce this probability of failure we have to design for a return period of $T_1 > T$.

One has to remember that for large values of T the probability of observing a flood $\{X > x_T\}$ in the next T years is about 63%. In fact, $\ln(1 - p_1) = T \ln(1 - 1/T)$. When T is large enough, a Taylor expansion leads to $\ln(1 - p_1) \approx T(-1/T) = -1$ and therefore $p_1 = 0.6321$. This means that a flood with a return period corresponding to 1000 years has a probability of occurrence of 63.21% in 1000 years! This is a high probability. We only have to remember that the return period is an average over a very long period of time.

Using equation (2.5) we can also determine the return period T to choose in order to obtain a probability of failure p_2 based on a life cycle of N years:

$$p_2 = 1 - \left(1 - \frac{1}{T}\right)^N \Rightarrow T = \frac{1}{1 - (1 - p_2)^{\frac{1}{N}}} \tag{2.6}$$

Example

N	p_2	T
100	0.63	100
100	0.20	450
100	0.10	950

If the objective is a probability of failure of 0.10 for a life cycle of $N = 100$ years, the structure must be designed for a high return period, equal to or greater than $T = 950$ years.

2.2.2 Probability that the First Event $\{X > x_T\}$ will Take Place after *N* Years

Let $N = 10$ years. In the above example we can make direct use of the fact that in the 10 first years the statistical trial should be a "failure" with probability $(1 - p)$: consequently, the probability of interest is equal to $(1 - 0.02)^{10} = 0.82$. The same result can be obtained by using the geometric distribution $\mathcal{G}(0.02)$. In a general manner the probability that the first event $\{X > x_T\}$ will have taken place after N years is equal to $(1 - p)^N$.

2.3 RISK

2.3.1 The Meaning of Risk

"Qu'est-ce qu'on risque?" is a popular expression in the French-speaking region of Switzerland which could be translated in English as "What's the risk?" – which means in fact that there is no risk at all! We mention this expression to illustrate that the word "risk" carries different meanings in various contexts, and some of the meanings are totally different. It seems thus necessary to explain this concept with more detail to eliminate any ambiguities and confusions regarding this term. Consider the following examples:

- There's a *risk* of rain later today.
- I'm taking a *risk* on a new mining stock.
- The *risk* of accident is particularly high this Easter week-end.
- Don't take any *risks*!
- Be careful, you run the *risk* of breaking a leg!
- Etc.

These expressions reflect both the notion of the *probability* of an event – and also the notion of the *consequence* of the event.

That being the case, two questions naturally arise:

- Is it possible to come up with a rigorous definition of risk that incorporates the concepts of both probability and consequence?
- Is it possible to quantify risk, or in other words to measure the level of risk in order to, for example decide whether it is necessary to take measures to reduce the risk to an acceptable level?

2.3.2 Expressing Risk

As we have just seen, the term risk involves two concepts: The first is the *probability* that a given event will take place, and the second is the *consequence* of this event, which could be either positive or negative. Following the example of several specialists in statistics as applied to hydraulic design (for example Bessis, 1984) we define risk mathematically as:

$$R = pc \tag{2.7}$$

This means that the risk R is expressed as the product of the probability p of the occurrence of an event of concern (in a particular interval of time, usually one year) and the cost c of damages (the consequences) that this event causes. Note that French-speaking specialists are more and more likely to use the term "hazard" to designate the event and "vulnerability" to express the consequences[5].

It is fair to ask why it is judicious to combine with a product the two terms *probability* and *consequence*? Two simple examples show, in a pragmatic fashion, the relevance of this practice:

- Let us suppose that a given event $\{X > x_T\}$ has an annual probability of occurrence of $p = 0.05$ (and therefore a corresponding return period of $T = 20$ years). This means that on average over a very long period, the event will occur every 20 years[6]. If each time this event causes \$100 000 worth of damages, the average annual cost of damages is \$5000. So here we have used the product $R = pc$, which defines the annual risk.
- This definition of risk also conforms to the standard practice of insurance companies. By adding a share of the administration costs and a contribution to the company profits, the insurer arrives at the premium to be billed to the policy-holder.

Equation (2.7) is only applicable in cases where we are concerned with a single threatening event. In the more general case of N threatening events we would use:

$$R = \sum_{i=1}^{N} p_i c_i \tag{2.8}$$

This relation is nothing more than the weighted sum corresponding to the expectation $\mathrm{E}[C]$ of the costs.

If the number of threatening events becomes infinite (for example, each of them is characterized by a continuous variable) this relation must

[5] See Dauphiné (2001) for example.

[6] Again, it should be noted that the concept of return period does not imply any regularity in calendar time: it is just an average.

take the form of an integral (the case of the expectation of a continuous random variable):

$$R = \int c f_C(c) dc = \mathrm{E}[C] \tag{2.9}$$

2.3.3 The Perception of Risk

The formal definition of risk, as presented above, allows for its direct quantification. We could very well stop there and agree that Equation (2.7) allows us to determine whether a particular risk is acceptable or not by setting a conventional level of acceptable risk. Figure 2.2 illustrates such an approach with an iso-risk curve of probability and consequence. This curve, which is an equilateral hyperbola, delineates the area of the acceptable zone.

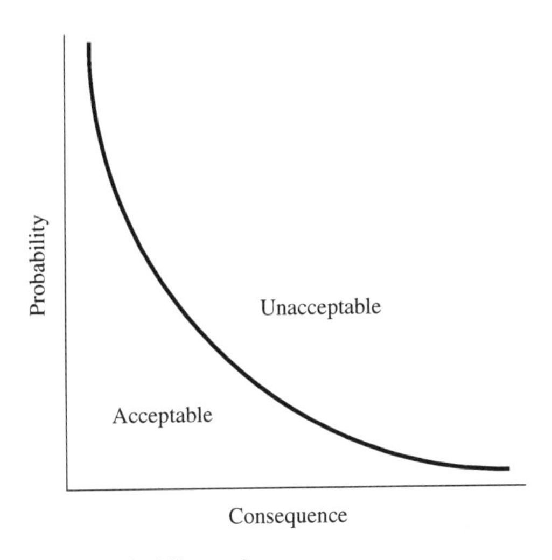

Fig. 2.2 Iso-risk curve of probability and consequence.

The choice of this "border" between acceptable and unacceptable risk is more a subjective and political decision than a scientific process. There are numerous factors that influence the degree of intensity of the perception of a risk, but here we mention only two – time and distance – to illustrate the problem (Dauphiné, 2001). Figure 2.3 shows that the intensity of the perception of the risk is inversely proportional to the factors of time and distance. Humans are known to forget past disasters (alas!), to the point that after a number of years, the perception of risk has become quite low. Conversely, if a disaster occurs within the geographic proximity of someone, he or she will experience it with great intensity.

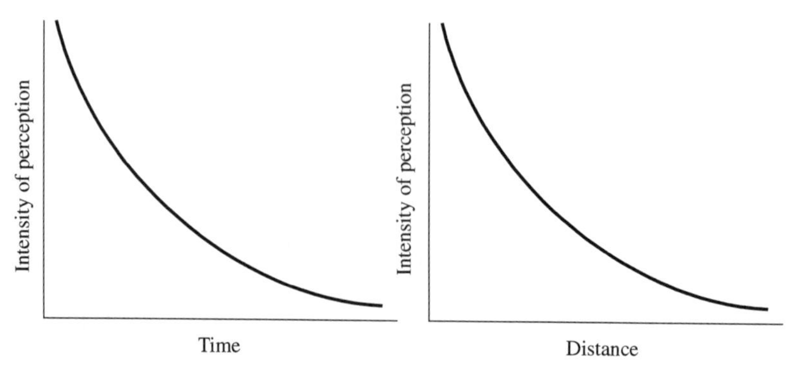

Fig. 2.3 Degree of the perception of risk as a function of time and distance (based on Dauphiné, 2001).

2.3.4 Particular Methods for Reducing Risk

The definition of risk written in Equation (2.7) is instructive. In essence, if we only set our safety objectives in terms of the probability p of the event from which we are trying to protect ourselves, or its corresponding return period T, we can immediately see that this probabilistic aspect is not the only relevant factor. It is actually possible to seek to decrease the costs c of damages through a whole series of legislative, organizational or constructive measures. This is the favored approach in many countries. The following list of possible measures is for the purpose of example but by no means exhaustive:

- land-use planning measures (restrictions on the use of land),
- prevention,
- forecasts,
- warnings,
- rescue organization,
- contingency plans,
- training,
- information,
- maintenance,
- constructive measures (for example, likely to avoid a chain reaction),
- crisis management,
- etc..

2.3.5 Types of Costs and the Economic Optimum

Let us imagine for a moment that the consequences of an event can be expressed solely in monetary terms[7]. Naturally, in this case risk R is also expressed as a sum of money. This makes it possible to determine the economic optimum of a protection system against the threatening event we are considering. To accomplish this, we have to consider the different costs, which – for comparison purposes – will be expressed on the same time scale:

- Cost of construction of a protection device, expressed in terms of its annual *amortization* U_1. This makes it necessary to make a financial calculation using the amortization period and the interest rate.
- Annual *operating cost* U_2. This amount includes costs for maintenance, personnel, data gathering, etc.
- Annual cost of *residual risk* U_3. This involves the risk R given the presence of a protection structure, computed by using the year as the reference period for the probability of the occurrence of the threatening event $\{X > x_T\}$.

It is possible to assess several protection variants to reach the total annual cost U, which is the sum of the costs of amortization, operations and residual risk:

$$U = U_1 + U_2 + U_3 \tag{2.10}$$

If the different variants examined do not differ except for the degree of intensity of the threatening event from which we are trying to protect ourselves, the economic optimum for reducing the risk can be determined by using as graph such as the one shown in Figure 2.4.

2.3.6 The Non-monetary Aspects of Damages

If the monetary aspects were the only cost to consider, the concept of risk $R = pc$ would be easy to apply. Unfortunately, such is not the case. When considering losses other than financial (human, social, political costs ...) the idea that one could find "the" optimal solution is an illusion. In fact, it's a good bet that such a solution simply does not exist. For example, how would it be possible to compare the cost of damages to personal property or real estate to the loss of human life? We could try to answer this by using the techniques of *multicriteria decision analysis,* which is often used in negotiations involving a number of partners with different viewpoints.

[7] We will come back to this important limitation later on.

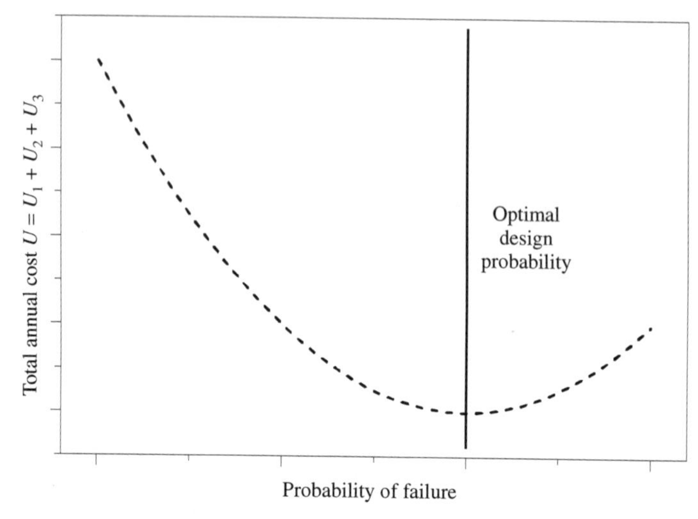

Fig. 2.4 Determining the economic optimum of protection (based on Sokolov *et al.*, 1976).

2.3.7 Approaches to Risk

Each country has its own particular approach for quantifying risk and applying the relevant legislation. Since none of the efforts that have been made on this question have been perfectly fixed and settled either in practice or legally, we will mention only one for the purpose of illustration.

In Switzerland for example, the Office fédéral de l'environnement, des forêts et du paysage[8] (OFEFP, 1991a) proposed that nine indicators for each particular type of damage be considered. The first group of these indicators describes damages i) to human and animal life, ii) to the environment and the basics of life. Only the last indicator deals with tangible assets.

Each of these indicators is then *standardized* in a *damages index* that varies from 0 to 1 ranging from a "light" through serious damage up to catastrophe.

Obviously, the choices in this scale involve a great deal of subjectivity. Certainly the process involves long discussions and negotiations among experts in the various fields concerned. But despite this shortcoming, such an index makes it possible to evaluate risk through a diagram such as the one in Figure 2.5.

[8] Today the Office fédéral de l'environnement (OFEV).

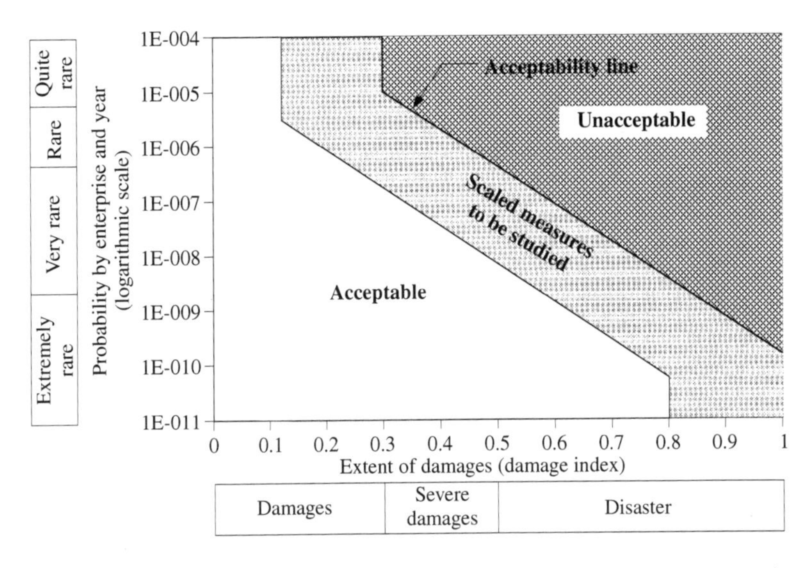

Fig. 2.5 Iso-risk diagram and categories of acceptability (based on OFEFP, 1991a).

Evaluating the scenarios corresponding to the different probabilities and damage indexes makes it possible to plot positions in an iso-risk diagram (cf. Fig. 2.5). Note that because the scales are logarithmic, the iso-risk lines are straight lines. On the right: a serious catastrophe is unacceptable under any conditions. Likewise on the left: accidents causing little damage are tolerable and are not considered in the directives of the OFEFP. The practice is to start with the dominant damage index among the nine indices. In cases where none of the indices is dominant, a synthesized index is used (OFEFP, 1991a).

2.4 UTILIZATION OF STANDARDS

The analysis described in the previous paragraphs is often complicated, time-consuming and expensive. That is why it is the current practice in most countries to remain content with ascertaining the probability of occurrence of a threatening event, and then proceed to protect against it by means of adequate construction.

The value of the design of a structure is thus determined solely on the basis of a frequency analysis. The corresponding threshold of probability is however generally defined by considering two additional criteria:

- the type and magnitude of the structure;
- the population density or the economic importance of the region.

Again, there is so much diversity in national practices that an *inventory* of these practices would be useless.

A Canadian Example

In Quebec (Canada), minimum safety standards for dams were laid out in the Dam Safety Act, which came into effect April 11, 2002. Section 21 stipulates that "*Subject to sections 23 and 24, the characteristics of every dam must ensure that it can withstand as a minimum the safety check flood described in the Table below that corresponds to the failure consequence category of the dam under section 17 and 18. However, if the dam failure consequences were assessed on the basis of a dam failure analysis, the category to be considered for the purpose of the table is the highest consequence category resulting from the examination of dam failure scenarios in flood conditions.*"

Table 2.3 Probability of recurrence with relation to level of consequences (Dam Safety Act, Quebec).

Consequence Category	Safety Check Flood (Recurrence Interval)
Very low or low	1 : 100 years
Moderate or High	1 : 1 000 years
Very high	1 : 10 000 years or ½ PMF
Severe	Probable maximum flood (PMF)

The level of consequences is determined based on the characteristics of the affected territory (see Table 2.3).

CHAPTER 3

Selecting and Checking Data Series

For the most part, the acquisition, preprocessing and archiving of hydrometeorological data falls outside the scope of this book. These aspects and especially the great diversity of measurement techniques (manual and automated) and methods for storing basic information (field books, analog recordings on paper, digital recordings, teletransmissions, etc.) lie within the area of basic hydrology or operational hydrology.

However, to carry out a frequency analysis, it is necessary to use the available data to build a *series of values* corresponding to a *sample* in the same sense that statisticians use these terms.

The construction of the series to be processed depends on the type of handling that to be carried out, or, conversely, the type of handling depends on the nature of the series of data available. In addition, for the statistical processing to be legitimate, the series under study must satisfy certain *criteria*.

This chapter uses several examples to suggest the steps to follow to build data series. The second part is dedicated to the validation of the series of values obtained.

To date, the issue of the *definition of the goals of the analysis,* suggested by Figure 1.2, has often been willfully ignored because it seemed so obvious. However, it cannot be stressed enough that it is extremely important to formulate the goals clearly and to adapt the steps in the analysis accordingly.

In this regard, one of the essential criteria is certainly *the spatial and temporal scales*: studying the behavior of floods in an urban micro-watershed (with a very short concentration time) with monthly rainfall data would not make any sense! The reverse is also true: it is useless to use rainfall data with a time step of a minute to study the Amazon watershed.

3.1 CHOICES OF EVENTS

3.1.1 Basic Information

In the analysis of data (rainfall or discharges), the available values themselves (which are the results of the measurements) are usually not of great interest, but rather the phenomenon or the *process*[9] that these values reflect. Let us consider the case of discharges measured continuously (which, in practical terms, is at a time step of 10 minutes in the example that follows, very much smaller than the lag time of the watershed). The process under study is expressed by the function $Q(t)$.

Let us suppose that for a given gauging station, the discharge rating curve is already known. Figure 3.1 shows a rating curve for the Corbassière River (Haute-Mentue, Switzerland).

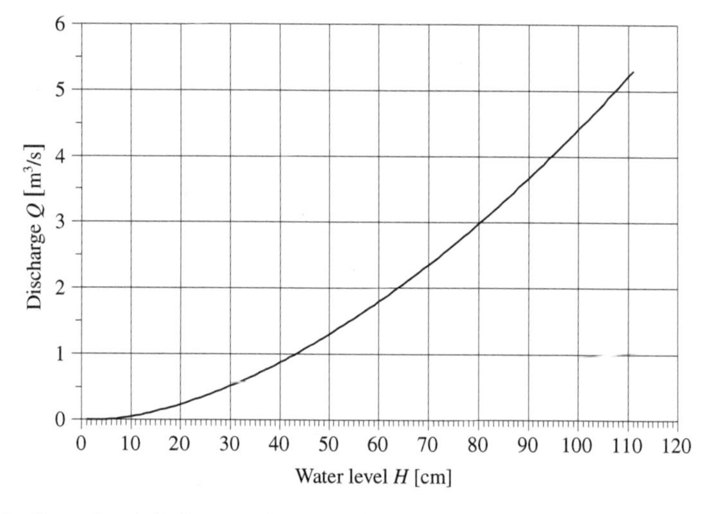

Fig. 3.1 Example of discharge rating curve for the Corbassière River (Haute-Mentue).

A continuous recording of the water depth $H(t)$ from a limnigraph (see Fig. 3.2) gives a good insight into this process and makes it possible to use a discharge rating curve expressed by the following equation:

$$Q(t) = f\{H(t)\} \tag{3.1}$$

to obtain $Q(t)$ (Fig. 3.3).

---

[9] A natural phenomenon marked by gradual changes that lead toward a particular result. (Merriam Webster)

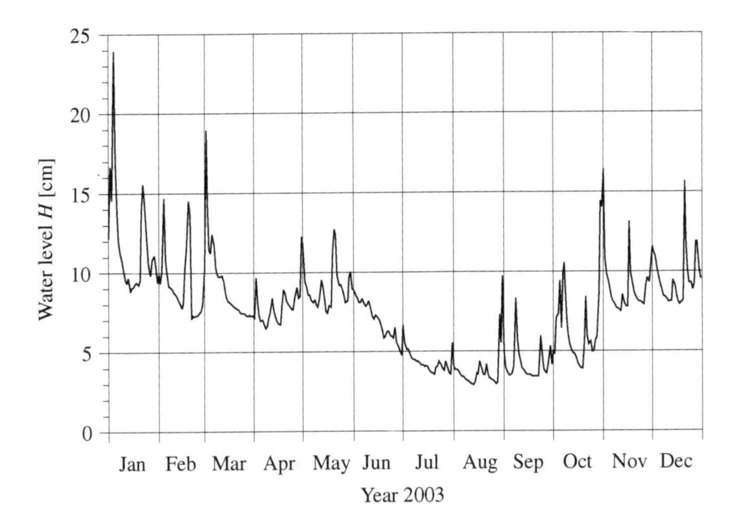

Fig. 3.2 Example of a limnigram of the Corbassière River (Haute-Mentue) in 2003.

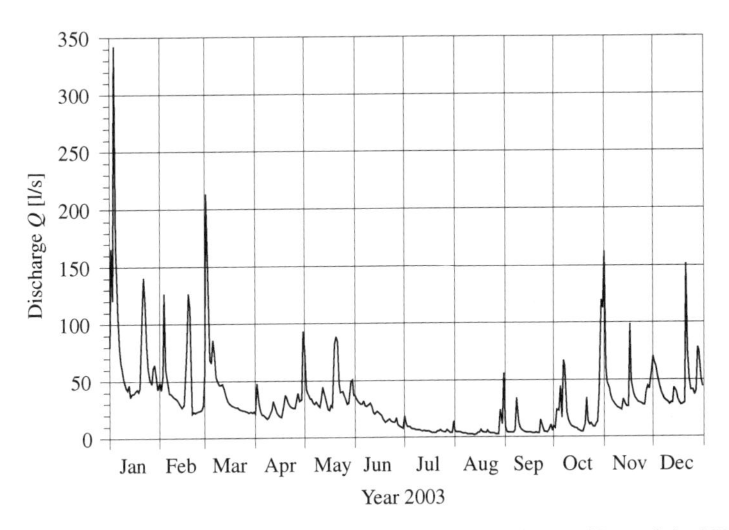

Fig. 3.3 Example of a hydrogram of the Corbassière River (Haute-Mentue) in 2003.

Very often, engineers do not have access to such comprehensive information (in the temporal sense). In this case they have no choice but to use the available data, for example in hydrological yearbooks. It should be noted that this practice makes it possible to benefit from the sometimes large amount of data preprocessing work done during the preparation of these publications; on the other hand, the relation between the time resolution of the available data and the nature of the problem to be solved may not be adequate.

Usually hydrological yearbooks publish the daily mean discharges, and less frequently the daily peak discharges.

It should be noted that the development of digital recording devices has made available pseudo-continuous information (which is to say, with time steps much smaller than the lag time of watersheds).

For example, in the case of rainfall measured with a float recording gauge, the information can be considered as pseudo-continuous: depending on the speed that the paper unrolls, it is theoretically possible to determine the quantities of rain $P(t)$ that have fallen in any period of time, even for very short time steps.

Real-time recording with a tipping-bucket rain-gauge also provides pseudo-continuous information: the capacity of the buckets (usually 0.2 [mm] or 0.1 [mm] of rainfall) is quite small compared to the quantities that are of interest for operational hydrologists.

In the case of rainfall, the engineer usually only has access to a time series of "aggregated values for x hours".[10] Usually, these are the depths of rainfall for 24 hours as published by meteorological services: in Switzerland, rainfall as published by MeteoSwiss corresponds to the measurements recorded every morning at 07:00 (GMT+1). For a smaller number of stations, rainfall series for 1 hour or for 10 minutes time steps are also available (in the case of the current ANETZ network of MeteoSwiss). The rainfall <u>*in*</u> x hours is estimated based on the rainfall data <u>*for*</u> x hours by applying a correction known as the *Weiss correction* (1964).

As shown in Figure 3.4, a measuring device with a constant time step leads to a *systematic* under-estimation of the rainfall intensities. Weiss analysis is meant to correct this bias and leads to the following equation:

$$F_{Weiss} = \frac{n}{n - 0.125} \qquad (3.2)$$

where n is the relation between the duration of the studied events and the measurement time step.

The experience of operational hydrologists has confirmed that the Weiss correction is perfectly valid.

---

[10] Following in the footsteps of Michel (1989) we will use the expression "rainfall of x hours" to indicate that the interval of time over which rainfall has been measured is fixed. As for the expression "rainfall in x hours," this will be reserved for the maximum quantity of precipitation over duration of x hours, where a moving window is considered through a rainfall event.

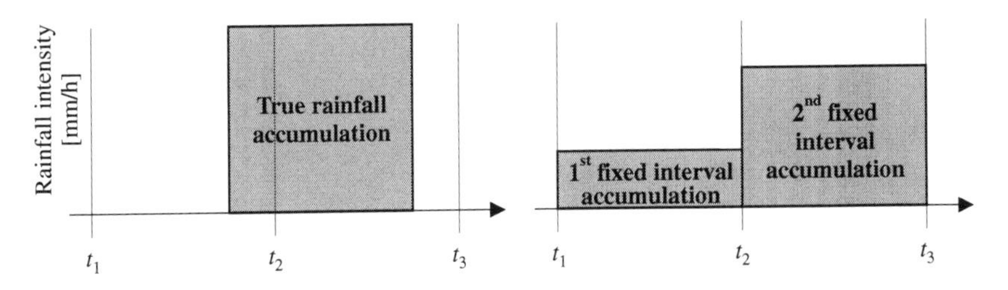

Fig. 3.4 Principle of Weiss correction.

We will use the term *basic time series* for the data available at this stage, which comes from the temporal sampling of the process under study. We have just seen that the characteristics of this series can be highly variable, depending on the type of information available.

3.1.2 Historical Information

In this book, we are interested mainly in historical data linked to discharge (rather than precipitation or temperature). Estimating discharges with large return periods requires a sufficiently large sample size. However the quantity of recorded hydrometric data at a particular site is often insufficient to allow for a reliable estimation. In this case the hydrologist is interested in finding additional information. For example, regional estimation methods have been developed to compensate for this lack of information for a particular site. Various studies have shown that the inclusion of historical information makes it possible to improve the accuracy of estimates of extreme floods. Historical information concerns floods that occurred before systematic discharge measurement became common.

3.1.3 Data Series of Characteristic Values

To undertake a frequency analysis, the first task of the hydrologist is to define the *characteristic variable* of interest (peak flood, monthly discharge, low water, etc.). This statement may seem trivial, but it often happens that analyses are performed in a mechanical fashion without this question even being asked.

A clear description of the process and its schematization makes it possible to determine the *characteristic variables* that need to be extracted from the measured values: these characteristic values form a subset or a derivative set of the process under study.

This means, for example, that in the case of the discharge in the Corbassière River (Haute-Mentue), the following characteristic variables can be taken into consideration:

- maximum discharge of each flow (peak discharge),
- flood volume,
- low-water discharge,
- mean annual discharge,
- mean monthly discharge,
- maximum annual mean daily discharge,
- etc.

Several techniques for differentiating the characteristic values begin by splitting the process into hydrological events (partitioning the rainstorms or the floods) before extracting the desired values from each event (maximum intensity in x minutes, peak discharge ...). The purpose of this splitting of events consists of obtaining the characteristic values of independent events, in such a way that they satisfy the common rules of sampling for the purpose of statistical inference. This procedure is also very useful for developing a particular type of hydrological model: discrete-event models.

It must be stressed that in this case, the *splitting* operation is far from trivial and requires precision. If we are looking at rainfall events for the purpose of simulating peak discharges, the criteria to be adopted will be quite different than for storage volumes.

The basic concepts of the sampling process and the associated terminology are shown in Figure 3.5.

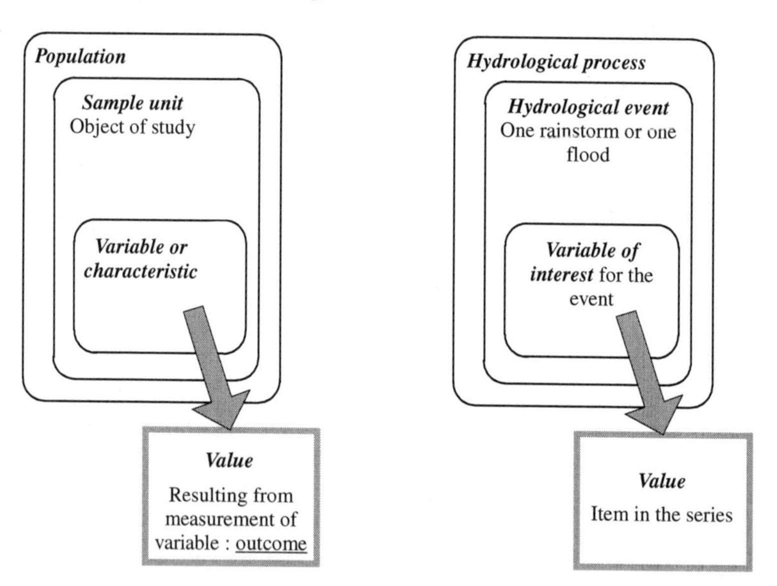

Fig. 3.5 "Statistical" and "hydrological" sampling terminology.

Figure 3.6 provides an example of the extraction of the characteristic values, the occurrence of peak floods, for two events. For each "flood event" only the maximum peak measurement is kept: the other local peaks are considered as *dependent* and are therefore ignored.

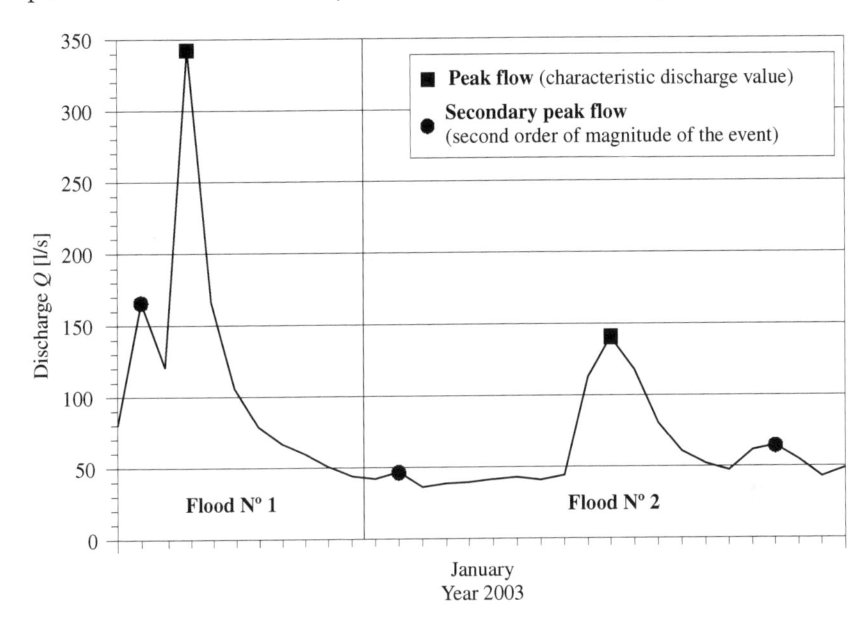

Fig. 3.6 Definition of characteristic values of events example: floods of the Corbassière River (Haute-Mentue), January 2003.

This example highlights the determining influence the method used for splitting the events has on the results of the extraction operation: the two splitted floods could just as easily be considered as building one single and longer flood. The issue of partitioning events is therefore a delicate task, and consequently, difficult to perform in an automatic way. In this context, the analysis goal and the related spatio-temporal scales are crucial criteria.

The next two paragraphs explain some criteria for partitioning rainstorms and floods.

3.1.4 Splitting of Rainstorms

A "rainstorm" is usually defined as the total quantity of rain associated with a well-defined meteorological disturbance. Thus the duration of a rainstorm could vary from a few minutes to many hours and affect an area ranging from a few square kilometers (a thunderstorm) to thousands of square kilometers (cyclonic rains). Ultimately, we can define a rainstorm as a continuous rain event that may have different intensities at different times.

A *hyetogram* is the representation in histogram form of the rain intensity as a function of time. It represents the derivative in a given point, in relation to time, of the accumulated precipitation curve. The important elements of the hyetogram are the time step Δt and the shape. The time step Δt is usually selected as a function of the time period represented; for example, an hourly time step for a weekly graph, a 10-minute time step for a daily graph, etc. The shape of the hyetogram is usually characteristic of the type of rainstorm and varies from one event to another.

The criterion for the continuity of a rain event varies depending on the watershed under study. In general, two rainstorms are considered as distinct events if: 1) the precipitation ΔH that falls during a time interval Δt between the events is less than a certain threshold and 2) this time interval itself is greater than a given value selected on the basis of the type of problem under study, for example the time of concentration of a watershed. When representing rainstorms in the form of hyetographs, the problem of how to split the rain events is summarized in Figure 3.7.

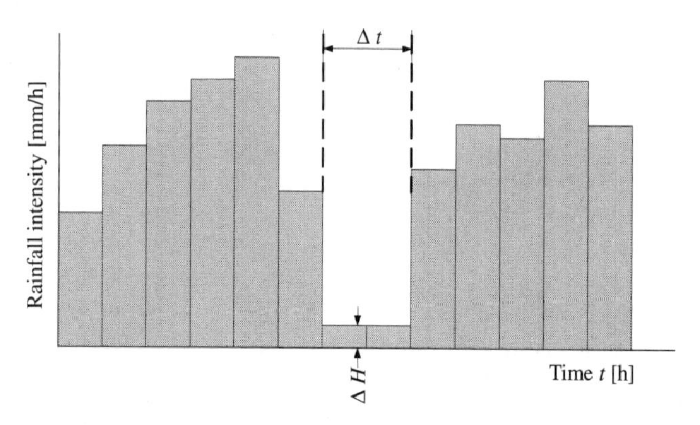

Fig. 3.7 Conditions for splitting two consecutive rainstorms (1) time period during which intensity i < threshold ΔH (for example 2 [mm/h]) and (2) Δt > time duration chosen on the basis of the problem under study (for example 1 hour).

3.1.5 Splitting of Floods

The shape of a particular flood is characteristic of the studied watershed (for example its size or geographic location). As a consequence it is very difficult to do the splitting of floods in an automatic way. Very often the split is done manually and depends on the experience of the hydrologist.

However several software have been developed in order to make this splitting automatic or semi-automatic. The CODEAU® software (De Souza *et al.*, 1994), for example, splits floods (or discharge events) on the basis of three parameters to be chosen:

- the entry threshold Q_E,
- the exit threshold Q_S,
- the time duration between two floods D_{sep}

The first step consists of examining the time series of the (observed) discharge (Fig. 3.8): a flood begins when the discharge reaches the entry threshold Q_E and ends when the discharge becomes smaller than the exit threshold Q_S.

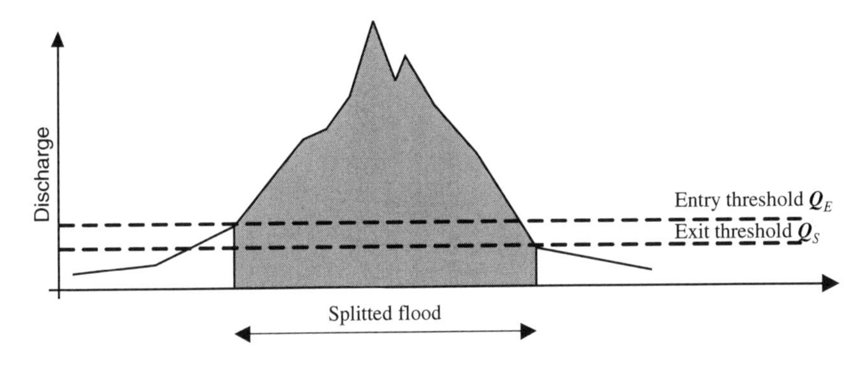

Fig. 3.8 First step in splitting floods: thresholds choice.

The second step is to verify the characteristics of the interflood episode, that is:

- if the duration between the two floods is greater than the chosen parameter D_{sep} the two floods are considered as distinct;
- if not, the floods remain grouped together.

After splitting, the criteria of peak flood discharge (Qp_{min} and Qp_{max}) and of flood duration (D_{min} and D_{max}) are checked again to decide whether to keep the flood in the sample.

There are other examples of splitting criteria. The Natural Environment Research Council (NERC, 1975, page 14) proposes a two-part rule:

- the minimum interflood discharge must be less than 2/3 of the peak discharge of the preceding flood;
- the time interval Δt separating the peaks of two successive floods must be a minimum of three times the duration of the mean rise of the hydrogram.

According to Beard (see Cunnane, 1989, page 11) two successive flood peaks must be separated by more than 5 days + ln A, where A is the surface area of the watershed in square miles.

3.2 TYPES OF SERIES

Figure 3.9 illustrates how by starting with a basic time series, it is possible to extract four types of time series that can be used for a frequency analysis:

- a complete duration series,
- annual extremum time series (sometimes called "annual maximum time series") the most commonly used in frequency hydrology),
- peak over threshold (POT) series (obtained from values above a given threshold),
- series of k greatest annual values (also called "inflatable-parameter series").

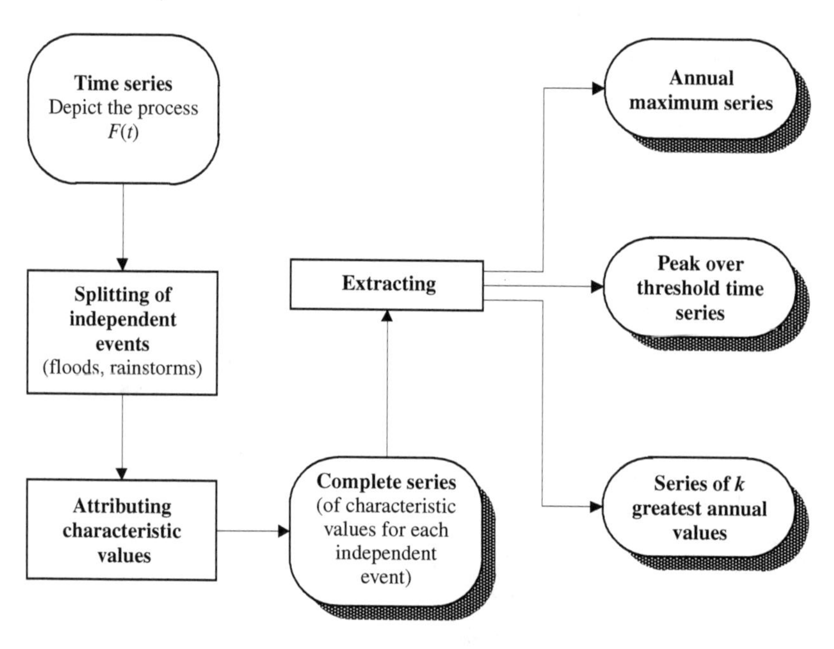

Fig. 3.9 The four types of series.

3.2.1 Complete Duration Series

At this stage in the analysis we have at our disposal the results of the measurements of the relevant characteristics for each independent event.

A series of events such as this is called a *complete duration series* (implies a complete series of the characteristic values of independent events). In our example, this involves a series of instantaneous peak discharges.

Various statistical techniques can be applied to check the independence of the successive events building the sample, some of which are described later in this chapter, such as:

- test of the nullity of the autocorrelation coefficient,
- the runs test.

If a possible autocorrelation (meaning a certain "internal" dependence) is detected as significant, this does not necessarily imply that every later computation will be compromised. First of all, it would be wise to check the data, as well as the procedure for splitting the events. On the other hand, there exist methods that make it possible to allow a certain degree of autocorrelation in the data (see for example Michel, 1989, page 100).

3.2.2 Annual Extremum Time Series

A complete series of events can contain a huge number of values. Using the complete series is sometimes unwieldy and in addition can produce a nonhomogeneous sample.

Based on the complete series, it is possible to extract a series of annual characteristic values (maximum, minimum, mean, etc.), such as an annual maximum series.

Practical frequency analysis shows that for time series longer than thirty years, it is valid to limit the analysis to annual extreme values. In fact, this is usually the recommended approach: it allows for direct use of a calibrated frequency model with an *annual* probability of occurrence. However, this way of proceeding can be tricky when we are dealing with extreme values. For years without severe floods we would include an ordinary flood in the sample. Conversely, for years with more than one severe flood, we would be lacking a value for one severe flood. The POT methodology, introduced to increase the size of the series, frees us from this problem.

3.2.3 Peak Over Threshold (POT) Time Series

When the sample size is too small, it is preferable to find a way to make better use of all the available information: we refer to this as a *comprehensive* analysis of information.

The first method we can use in this case consists of keeping from the complete series only the independent events with values above a given threshold x_0 that is chosen arbitrarily. Such a truncated series is called a *peak over threshold series* or *partial duration series*. Using the frequency model resulting from this type of series is not straightforward: some particular developments are required (see Chapter 8).

In France in particular, this type of analysis is known as the *méthode du renouvellement*, or renewal processes method (see for example Miquel, 1984). For the choice of threshold, Miquel (1984) advocates an iterative technique that consists of repeating the full set of computations for different thresholds, in order to ensure a degree of stability in the results.

Experience has shown that in order to ensure a "quality" of estimation equal to or greater than that obtained by using an annual extremum series, we must have at our disposal a mean number of events ≥ 1.65, or, in other words, more than 1.65 events per year, on average (NERC, 1975) (see boxed text below). In practice, a series with a mean number of events per year on the order of 5 or 6 would be used, still respecting the criteria of independence between successive events.

Guillot (1994) proposed considering the 5% largest values for fitting the *upper tail* of the distribution of maximum daily rainfall (in connection with the so-called Gradex method). This approach results in about 18 annual values being considered. Several statistical methods for selecting a threshold are discussed in Coles (2001). They are based mainly on the idea that the sample should present a degree of homogeneity: we look for the threshold above which we find "extreme" events, for which GPD models are appropriate (see § 4.2).

Comparing the efficiency of two procedures can be done using a mathematical approach or a simulation. As an example, we have simulated (see also the Parametric Bootstrap in Chapter 7) a large number of samples of size 30 following a Gumbel distribution (suitable for a series of annual maxima), estimated its parameters and estimated the quantile corresponding to a 30 year return period. Then the mean and standard deviation of these quantiles were computed. At the same time we applied the same procedure using a POT series (which follows an exponential distribution, as we will see later in Chapter 8) for different parameters λ. This made it possible to compare the standard deviations obtained with both procedures.

Figure 3.10 below shows the ratio of the standard deviation (or standard error) of the quantiles obtained with the POT series (as a function of λ) to the standard deviation obtained using an annual maximum time series. Although this direct approach is a little simplistic, it allows us to conclude that we should choose a parameter of $\lambda > 2$ for a POT series in order to obtain an efficiency equal to that obtained with the annual maxima method.

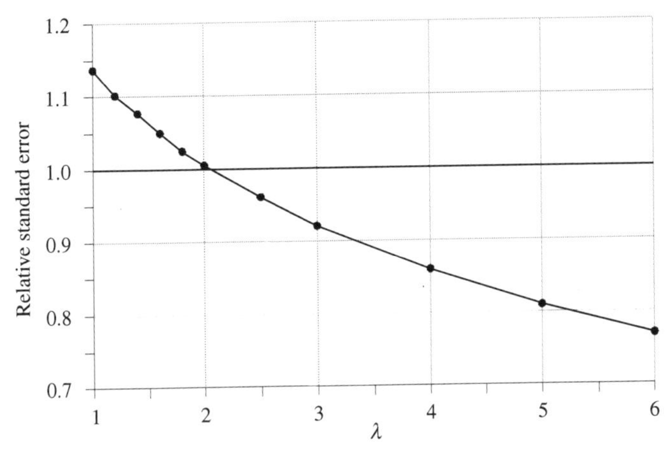

Fig. 3.10 Ratio of the standard deviation of the quantiles obtained with the POT series to the standard deviation obtained using an annual maximum time series.

3.2.4 Series of *k* Greatest Annual Values ("Inflatable-parameter" Series[11])

The second method that is often applied when the sample size of annual maximums is too small consists of extracting, for each year, the *k* largest values from the complete series. This is called a *series of k greatest annual values* or in more common terms, an "inflated" series.

Again, the correct use of this type of times series in a frequency analysis requires some further development (see Chapter 8).

However, this technique does not yet have a well-established statistical basis and should be considered as a second-best solution: when at all possible, it is preferable to use a POT series.

3.2.5 Nonsystematic Time Series

There are several ways for classifying non-contemporary flood information, including classification based on its roots (Ouarda *et al.*, 1998) or on its chronology (Baker, 1987).

Here we will discuss chronological classification as advocated by Naulet (2002). This is based on a split into three periods: prehistoric, historic and contemporary. In terms of hydrology, the *prehistoric period* (the scientific domain of the *paleohydrologist*) is studied by using techniques based on geological, geochemical and biological indicators. Data produced by paleohydrology is censored data[12], but in some cases

---

[11] The corresponding term in French (*série gonflée*) is commonly used in Switzerland.

[12] In statistics, the definition of censored data comes from survival analysis. Inherently the phenomenon of censoring appears when there is information about the *survival time* of an individual (for example, that a person lived past a certain date), but the precise survival time is not known.

their intensity can be estimated and the date of occurrence can be known with a degree of certainty.

The *historic period* is basically determined through archives, and the technique mentioned above can sometimes be used for historic data as well. The historic period is split into two sub-periods: the older one contains noteworthy local observations (the data is higher than an observed threshold), and the more recent contains regular observations that resulted from the creation of flood warning services (eg., the late XIXth century in France). These sources of data usually involve daily records of water levels measured at regular intervals (one to three measurements per day). The number of measurements could sometimes increase during periods of flood.

The *contemporary period* began with the installation of gauging stations that provided a continuous discharge series by means of rating curves. The problem of extrapolating the rating curve for extreme events means that we should also consider oral or written accounts from the period. Figure 3.11 (adapted from Naulet, 2002) summarizes the chronological classification of historical data.

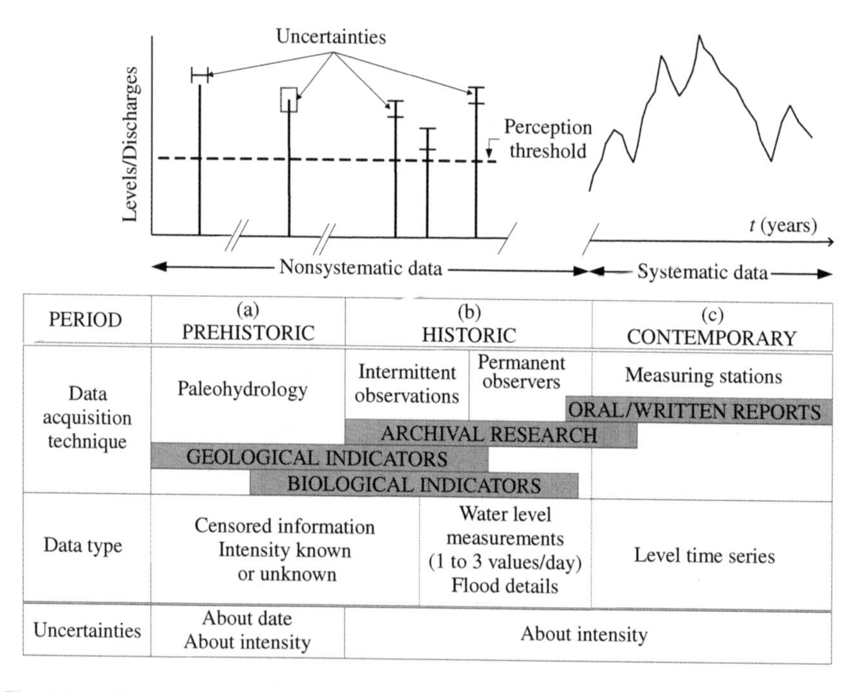

Fig. 3.11 Chronological classification of sources of information related to floods (adapted from Naulet, 2002).

Naulet (2002) explains in detail the method for collecting three types of flood data (prehistoric, historic and contemporary), and also described the problems involved in using prehistoric and historic flood data.

In frequency analysis, statistical processing is determined, if at all possible, by the nature of the data (censored or uncensored). The statistical processing of paleohydrological and historical information is similar.

Once the discharges have been reconstructed, these data represent POT or censored samples comprising:

1. discrete series of discharges of known peak intensity that surpassed a given threshold;
2. discrete series of values of known discharge intensities classified into several categories:
   - lower than a threshold during a given time period. If data is of the historical type, it could be based on, for example, the buildings that were not inundated. In this case, we are talking about a threshold of perception;
   - greater than a given threshold during a given period of time. In the case of paleohydrological data, this involves paleohydrological limits of exceedance;
   - included in an interval during a given period of time. It should be noted that this interval is longer than the threshold of perception.

3.3 TYPES OF ERRORS

The construction of a series of values building a sample in the statistical sense is a long process scattered with pitfalls, during which various different kinds of errors are likely to occur.

Errors can be committed during any of the four steps in the classical procedure, namely; measurement, data transmission, data storage and data processing (preprocessing and analysis). Therefore it is imperative, before using any data series, to pay attention to the quality and representativeness by applying various techniques that are usually either statistical or graphic-based (Musy and Higy, 2011, Chapter 9).

A measurement error is defined as the difference between the true value (which is the ideal we always seek, but which is in principle never knowable) and the measured value. It is useful here to distinguish two types of errors (both for understanding and in order to treat them): *random errors* and *systematic errors*.

Random errors (or accidental errors) — This type of error affects the accuracy of data and is not correlated. Such errors can result from a number and variety of causes, usually unknown, and affect each individual measurement differently. Usually these errors are considered to be the results of a normal random variable with zero mean and with variance σ^2. As these errors are inevitable, it is necessary to estimate their amplitude in order to include them in the final assessment of uncertainty. As far as possible, one should select measuring techniques that produce the smallest possible random errors.

Bias or systematic errors — This type of error affects the reliability of the data. We also use the term inconsistency in this case. Supposing that the measurements are not affected by any random errors, then the difference between the true value and the measured value, if there is a difference, is a systematic error. In most instances, systematic errors are due to the fact that the measurement instruments are not perfectly calibrated (equipment error, including especially measuring instrument drift, changing the observer...) or to an external phenomenon that disrupts the measurement.

3.4 ERROR IDENTIFICATION AND MEASUREMENT CORRECTION

Depending on the nature of the errors discovered or suspected, identification of the latter requires different techniques and methods, which are usually:

- "In situ" techniques which consist of checking on-site the way the data were organized, processed, and/or transformed in the field at the point the measurement was taken.
- Office investigations involve verifying the chain of measurement/data processing at each step in the process, following the same methodology used for data series submitted for data checks and/or publication.
- Statistical analyses using specific statistical software to bring to light certain errors or inconsistencies. These techniques are efficient and are used mostly in professional practice. They are based on specific assumptions that are well worth knowing.

3.4.1 Premises of Statistical Analysis

Statistical computations are based on a certain number of assumptions that need to be verified. Among these assumptions:

The measurements reflect the true values – This assumption is unfortunately almost never achieved in practice, due either to systematic or random errors.

The data are consistent – No modifications in the internal conditions of the system took place during the observation period (raingauge position, observation procedures, same observer...).

The data series is stationary – The properties of the statistical distribution underlying the phenomenon (mean, variance or higher-order moments) do not vary over time.

The data is homogeneous – A data series is considered as non-homogenous when:

- It results from measurements of a phenomenon with characteristics that changed during the period of measurement; the phenomenon is then non-stationary (for example: climatic variations, changes in discharge regime due to deforestation, reforestation or construction of a dam). Signs of seemingly non-stationary behavior can also be observed if the electronics integrated in the measuring device undergo a temporal shift, the observer is changed, or the readout device is moved.
- It reflects two or more different phenomena. The regime of a river downstream from the confluence of two sub-watersheds with very different hydrological behaviors is a good example of this lack of homogeneity.

The concept of homogeneity therefore implies stationarity.

The data series is random and simple – The random and simple characteristics of a series of observations are a fundamental assumption of statistical analysis. A *random sample* means that all the elements in a population have the same probability of being sampled. A *simple sample* means that the sampling of one element does not influence the probability of occurrence of the following element in the sample unit. In other words, if all the observation data in a series are from the same population and are independent of each other, it is a *simple random* sample. There can be many reasons, sometimes simultaneous, for the fact that the simple random characteristics cannot be verified. The causes fall into one of two categories: either autocorrelation defaults (non-random characteristic of the series) or stationarity defaults of the process (long-term drift and cyclic drift).

The series must be sufficiently long – The length of the series has an influence on sampling errors, especially on computation of higher-order statistical moments and therefore on tests of their reliability. In addition, a sufficiently long series is the only way to ensure good representativeness of the climatological process.

3.5 INTRODUCTION TO STATISTICAL TESTS

3.5.1 The Steps in a Test

This example is adapted from Saporta (1990). Measurements taken over a number of years made it possible to establish that the annual depth of rainfall in Beauce (France) in [mm] follows a normal distribution $N(\mu, \sigma^2)$ with $\mu = 600$ and $\sigma^2 = 100^2 = 10\ 000$. Some contractors dubbed rainmakers claimed that they could increase the average depth of rainfall by 50 [mm] simply by seeding the clouds with silver iodide, and that this would increase the agricultural yield of this breadbasket region of France (see Charles Péguy). Their process was tested between 1951 and 1959, and the rainfalls recorded for these years are shown in Table 3.1.

Table 3.1 Annual depth of rainfall in Beauce (France) in [mm] from 1951 to 1959.

Year	1951	1952	1953	1954	1955	1956	1957	1958	1959
Annual rainfall [mm]	510	614	780	512	501	534	603	788	650

What can we conclude from this data? There were two possible contradictory assumptions: either cloud seeding had no effect, or it effectively increased the mean rainfall depth by 50 [mm]. These assumptions can be formalized as follows:

If μ denotes the expected value of the random variable X "annual depth of rainfall," we can assume the following hypotheses:

$$\begin{cases} H_0 : \mu = 600\ [mm] \\ H_1 : \mu = 650\ [mm] \end{cases} \quad (3.3)$$

where H_0 is the null hypothesis and H_1 is the alternative hypothesis.

The farmers, reluctant to opt for the inevitably expensive process that the rainmakers offered, leaned towards the H_0 hypothesis, and so the experiment had to convince them otherwise; in other words, the experiment had to produce results that would contradict the validity of the null hypothesis.

They chose $\alpha = 0.05$ as the probability level, meaning they were ready to accept H_1 if the obtained results fell into the category of an unlikely event that has only 5 out of 100 chances of occurring. In other words, they implicitly accepted that rare events can occur without calling into question the validity of the null hypothesis; by doing this, they assumed the risk of being mistaken 5 out of 100 times in the case that the rare events actually occurred.

How to decide? Since the goal was to "test" the theoretical mean μ, it makes sense to look at the sample mean $\bar{x}$ (the average value of observations), which provides the most information about μ. $\bar{x}$ is called the *"decision variable."*

If H_0 is true, as experience had shown for $N = 9$ years, $\bar{x}$ must follow a normal distribution:

$$N\left(\mu, \frac{\sigma^2}{n}\right) = N\left(600, \frac{10\ 000}{9}\right) \tag{3.4}$$

In principle, high values of $\bar{x}$ are improbable and we use as a decision rule the following: if $\bar{x}$ is too large, i.e. if $\bar{x}$ is greater than a threshold k which has only 5 chances out of 100 of being exceeded, we opt for H_1 with a 0.05 probability of being false. If $\bar{x} < k$, we cannot reject H_0 for lack of evidence. k is called the *critical value.*

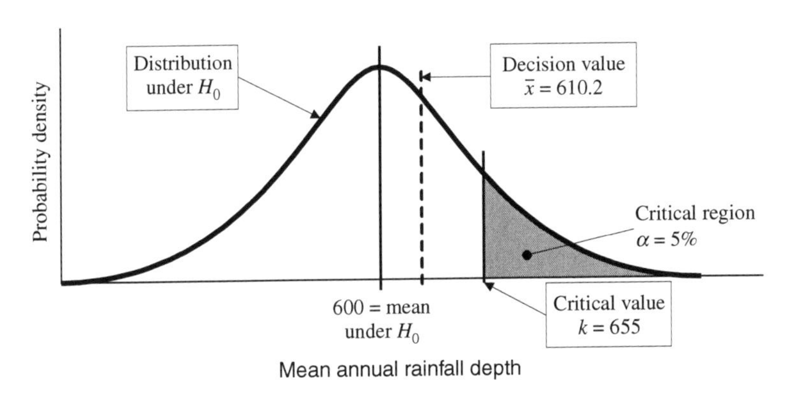

Fig. 3.12 Decision rules for the test under study.

It is easy to compute the critical value by using normal distribution tables. This gives us:

$$k = 600 + \frac{100}{3}1.64 = 655\ [\text{mm}] \tag{3.5}$$

The decision rule is thus the following:

- if $\bar{x} > 655$, reject H_0 and accept H_1;
- if $\bar{x} < 655$, keep H_0.

The events ensemble $\{\bar{x} > 655\}$ is called the critical region or rejection region for H_0. The complementary ensemble $\{\bar{x} \leq 655\}$ is called the non-rejection region of H_0. Figure 3.12 illustrates the concept of decision variable, critical value and critical region. Yet the recorded data indicate that $\bar{x} = 610.2$ [mm]. The conclusion was thus to keep H_0; that cloud-seeding had no noticeable effect on the amount of rainfall: the observed values could have occurred by chance in the absence of any silver iodide. However, nothing says that not rejecting H_0 protects us from mistakes: the rainmakers may have been right, but this fact was not detected.

There were in fact two ways to be wrong: to believe the rainmakers, whereas they had nothing to do with the results obtained (probability $\alpha = 0.05$); or not to believe the rainmakers, whereas their method was good and only randomness (unfortunately for them) due to the low number of observations, produced results that were not sufficient to convince the farmers.

If we suppose that the rainmakers were right, then an error is made each time that $\bar{x}$ takes a value less than 655 [mm], *i.e.*, with a probability of:

$$\beta = \Pr\left(Z < \frac{655 - 650}{100/3}\right) = \Pr(Z < 0.15) = 0.56 \text{ where } Z \sim N(0, 1) \quad (3.6)$$

which is considerable.

α is called type I error (the probability of choosing H_1 when H_0 is actually true) (the occurrence is 5% in this example); β is type II error (the probability of keeping H_0 when H_1 is true) (corresponding to 56% in this example).

These errors correspond to different errors in practice; thus, in the example of the rainmakers, the type I error consists of buying a cloud-seeding process that is worthless; the type II error leads to losing the opportunity to increase the amount of rain and perhaps produce more abundant harvests. In the practice of statistical testing, it is the rule to use α as the preferred given (for example the current values are 0.05, 0.01 or 0.1) as a function of type I error, which gives the primary role to H_0.

As α is fixed, β will be determined by computation (this is only possible if we know the probability distributions under H_1). However it should be noted that β varies in the opposite direction of α. If we want to lower α, a type I risk of error, we increase the 1-α probability of accepting H_0 when H_0 is actually true; but on top of this it results in a stricter decision rule that means not discarding H_0 except in extremely rare cases and therefore very often keeping H_0 by mistake.

In the case of this example, the statistical test would involve the following five steps:

1. Formulating and selecting H_0 and H_1.
2. Determining the decision variable.
3. Computing the critical value and the critical region relative to α.
4. Computing an experimental value for the decision variable.
5. Conclusion: reject or keep H_0.

3.5.2 Main Categories of Statistical Tests

Tests are classified either according to their objective (goal) or their mathematical properties.

Tests According to their Mathematical Properties

A test is considered to be parametric if it is based on a well-defined assumption regarding the underlying population. In most cases, these tests are based on the fact that the population follows a normal distribution and consequently assume the existence of a random variable of reference X distributed according to a normal distribution. The question is to find out whether the results remain valid even if the distribution of X is not normal: if the results are valid we say that the test under study is robust. The robustness of a test in relation to a given model corresponds to its ability to remain relatively insensitive to certain modifications of the model. A test is known as nonparametric if it does not use precise parameters or assumed hypotheses concerning the underlying distribution.

In the next chapter, we will indicate the parametric or nonparametric nature of the various tests by means of a boxed notation in the margin as following:

- Parametric test **P**
- Nonparametric test

Tests According to their Objective

These tests are usually classified into five main groups and contain most of the statistical tests generally used in hydrology.

1. *Conformity test*: comparison of a characteristic of a sample to a reference value in order to verify whether the corresponding characteristic in the population can be considered as equal to the reference value. For example, H_0: $\mu = \mu_0$; μ_0 is the reference value (or norm), μ is the mean, which is unknown, of the population.
2. *Homogeneity test or sample comparison test*: Given two samples of size n_1 and n_2, can we accept that they are issued independently from the same population? This problem can be formulated mathematically as follows: the first sample is issued from the random variable X_1 with distribution function $F_1(x)$ while the second sample is a realization of random variable X_2 with distribution function $F_2(x)$.

$$\begin{cases} H_0 : F_1(x) = F_2(x) \\ H_1 : F_1(x) \neq F_2(x) \end{cases} \tag{3.7}$$

The choice of H_0 depends on practical considerations because $F_1(x) \neq F_2(x)$ is too vague to obtain a critical region. In practice, we would be satisfied to verify the equality of the expected values and the variances of X_1 and X_2, by using the sample means $\bar{x}_1$, $\bar{x}_2$ and sample variances s_1^2, s_2^2 of the two samples.

3. *Goodness-of-fit tests*: Also called adequation tests, the purpose of these tests is to check if a given sample can be considered as coming from a specific parent population. In fact, this type of test comes down to carrying out a series of conformity tests on all the characteristics used for qualifying both the population and the sample. Usually this last approach is eliminated in favor of tests designed specifically for the purpose, such as the chi-square test, Kolmogorov-Smirnov test, etc.
4. *Autocorrelation test*: this type of test verifies whether there is a statistical dependence (due to proximity in time, for example) between the chronological data in a series of observations.

   The *autocorrelation* ρ_k with lag k of a stationary time series is expressed by the following equation:

$$\rho_k = \frac{\gamma_k}{\gamma_0} = \frac{\text{Cov}(X_t, X_{t+k})}{\text{Var}(X_t)} \tag{3.8}$$

   The *sample autocovariance* $\gamma_k = \text{Cov}(X_t, X_{t+k})$ is computed by means of a series of n observations $x_1, x_2, \ldots, x_n$ using the following equation:

$$\hat{\gamma}_k = \frac{1}{n-k} \cdot \sum_{t=1}^{n-k} (x_t - \bar{x})\,(x_{t+k} - \bar{x}) \tag{3.9}$$

   Autocorrelation is a measure of the persistence of a phenomenon.
5. *Stationarity test*: This category of tests includes trend tests, able to highlight a slow drift in the process, and rupture tests which seek to identify abrupt changes in the mean from a given date.

Tests Based on the Type of Information

In hydrology, different situations can occur as a result of specific hydrological situations. As a result, it is sometimes necessary to check a single type of data (rainfall, temperature, evaporation) at the local scale (where the measurements were taken) or at the regional scale (a watershed with a number of measurement sites). It is also sometimes desirable to check the quality of several types of data (e.g. rainfall-discharge, temperature-wind velocity...) at a local or regional scale. There are various data checks for this, including both numerical tests (strictly statistical) and graphic-based tests (a more hydrological practice), which can be classified into four main groups depending on the spatial scale and the number of parameters involved: one parameter – local scale; one parameter – regional scale; several parameters – local scale; several parameters – regional scale.

Data checking: example of the Vispa River discharge

The various tests described below will be illustrated using the following data representing the annual peak discharge in [m^3/s] of the Vispa River at the town of Visp (Switzerland) from 1922 to 1996 (Table 3.2). The particular characteristic of this discharge series is that the river underwent an anthropogenic change in 1964, when the Mattmark dam was built upstream from the measurement point.

Table 3.2 Annual peak discharge in [m^3/s] of the Vispa River at Visp from 1922 to 1996.

Year	Annual *Qp* [m^3/s]	Year	Annual *Qp* [m^3/s]	Year	Annual *Qp* [m^3/s]
1922	240	1947	210	1972	140
1923	171	1948	375	1973	115
1924	186	1949	175	1974	87
1925	158	1950	175	1975	105
1926	138	1951	185	1976	92
1927	179	1952	140	1977	88
1928	200	1953	165	1978	143
1929	179	1954	240	1979	89
1930	162	1955	145	1980	100
1931	234	1956	155	1981	168
1932	148	1957	230	1982	120
1933	177	1958	270	1983	123
1934	199	1959	135	1984	99
1935	240	1960	160	1985	89
1936	170	1961	205	1986	125
1937	145	1962	140	1987	285
1938	210	1963	150	1988	105
1939	250	1964	125	1989	110
1940	145	1965	115	1990	110
1941	160	1966	100	1991	115
1942	150	1967	85	1992	110
1943	260	1968	76	1993	330
1944	235	1969	110	1994	55
1945	245	1970	94	1995	63
1946	155	1971	150	1996	49

Figure 3.13 presents these data as a time series. The straight lines indicate the average discharges before and after the construction of the dam. Table 3.3 summarizes the main statistical characteristics of the two sub-series before and after the dam was built, as well as those of the complete series.

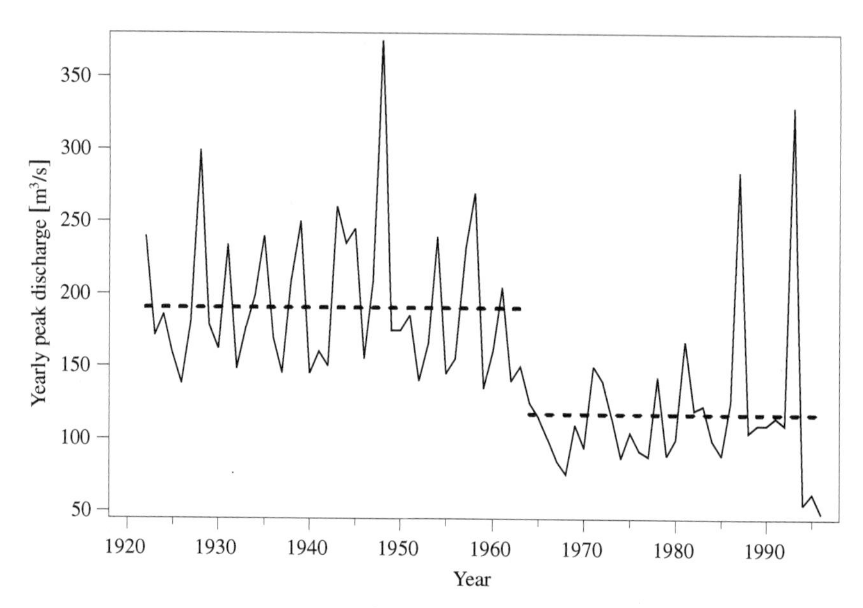

Fig. 3.13 Annual peak discharge [m³/s] of the Vispa River at Visp from 1922 to 1996. In 1964 the construction of the Mattmark dam induced a visible modification of the discharge regime.

Table 3.3 Main statistical characteristics of the two sub-series (before and after construction of the dam).

Series	size	Mean	median
1922-1963	$n_1 = 42$	$\bar{x} = 190.26$	$\tilde{x} = 176$
1964-1996	$n_2 = 33$	$\bar{y} = 117.27$	$\tilde{y} = 110$
1922-1996	$n = 75$	$\bar{z} = 158.14$	$\tilde{z} = 150$
Series	**standard-error**	**asymmetry**	**kurtosis**
1922-1963	$s_x = 48.52$	$\gamma_x = 1.47$	$\kappa_x = 2.88$
1964-1996	$s_y = 55.51$	$\gamma_y = 2.47$	$\kappa_y = 6.54$
1922-1996	$s_z = 62.99$	$\gamma_z = 0.90$	$\kappa_z = 0.96$

3.6 TEST OF ONE PARAMETER, LOCAL SCALE

In this example, we are testing the peak discharge series presented above (1 parameter) at the point measurement at Visp (local scale).

3.6.1 Conformity Tests

Conformity tests compare the mean or the variance of a sample to the mean or the variance of a theoretical distribution (of the population from which the sample is drawn). Two tests are used to check the conformity of the mean value depending on whether the variance is known or must be estimated, respectively the *z test* and the *Student test* (also known as the *t test*). (For more details on basic statistical tests, the reader is referred to a book of general statistics applied to hydrology, such as Helsel and Hirsch, 2002). These two tests are parametric: they assume that the population follows a normal distribution.

Mean

Example

We would like to test whether the mean peak discharge of the Vispa River at Visp for the period before the dam was built is equal to 200 [m^3/s]. For all the tests below, the selected significance level is $\alpha = 5\%$. Thus we have:

P

$$\begin{cases} H_0 : \mu_x = 200 \\ H_1 : \mu_x \neq 200 \end{cases} \tag{3.10}$$

Since the variance is not known, we will use Student's t test. The score is given by:

$$t_{obs} = \frac{\sqrt{n} \cdot (\bar{x} - 200)}{s_x} = -1.30 \tag{3.11}$$

As the test is bilateral, the critical value is given by the 97.5% quantile of the Student t distribution with (n-1) = 41 degrees of freedom. We have $q_{t_{41}}(97.5) = 2.01$ (see Tables of the Student distribution) and as a result, $|-1.3| < 2.01$. Therefore we can reject the null hypothesis that the mean peak discharge is equal to 200 [m^3/s].

Variance

The goal is to test if the variance is equal to a norm (written s_0^2). Determining the conformity of the variance is done based on the sampling distribution of the difference $s^2 - s_0^2$ in comparison to the variance ratio. The discriminant function follows a chi-square distribution. The procedure for this test is summarized in Table 3.4.

P

Example

A hydrologist assures you that the peak discharge variance of the Vispa River at Visp from 1922 to 1963 was 1600 [m^6/s^2]. Your own experience as a hydrologist leads you to think that the variance was actually greater than your colleague reports.

Table 3.4 Procedure for testing conformity of variance.

Step			
H_0	$\sigma^2 = s_0^2$	$\sigma^2 = s_0^2$	$\sigma^2 = s_0^2$
H_1	$\sigma^2 \neq s_0^2$	$\sigma^2 < s_0^2$	$\sigma^2 > s_0^2$
Discriminant function	$\chi^2_{obs} = \frac{n \cdot s^2}{s_0^2}$, follows a chi-square distribution with $(n - 1)$ degrees of freedom		
Non rejection domain of H_0	$\chi^2_{\alpha/2} < \chi^2_{obs} < \chi^2_{1-\alpha/2}$	$\chi^2_{obs} > \chi^2_{\alpha}$	$\chi^2_{obs} < \chi^2_{1-\alpha}$

In this case, the null and alternative hypotheses are:

$$\begin{cases} H_0 : \sigma_x^2 = 1600 \\ H_1 : \sigma_x^2 > 1600 \end{cases}$$

The discriminant function is given by: $\chi^2_{obs} = \frac{ns^2}{s_0^2} = 61.80$

As the test is right unilateral, the quantile to consider is the 95% quantile. A reading of the numerical tables tells us that the value of the quantile $q_{\chi^2_{41}}(95\%) = 56.94$, which is less than 61.80, and consequently we can reject the null hypothesis. Your intuition was right!

Wilcoxon Test

The most classical test in this case is the nonparametric test of conformity of the mean by *Wilcoxon* (Helsel and Hirsch, 2002, page 142.) To illustrate it, we will use the same example as in the parametric case above – Student's test – namely:

NP

$$\begin{cases} H_0 : \mu_x = 200 \\ H_1 : \mu_x \neq 200 \end{cases} \quad (3.12)$$

The Wilcoxon score for a single sample is expressed by the equation:

$$W^+ = \text{sign}(x_1\text{-norm})R^+(x_1\text{-norm}) + \ldots + \text{sign}(x_n\text{-norm})R^+(x_n\text{-norm}) \quad (3.13)$$

where R^+ is the signed rank or the rank corresponding to the absolute value of the signed observation, and where sign (u) is equal to 1 if $u > 0$, otherwise it is equal to zero.

Example

In the case we have been studying regarding the discharge at Vispa, $W^+ = 497$. For a sample size larger than 15, the following normal approximation is possible:

$W^+ \sim N\left(\frac{n(n+1)}{4}, \frac{n(n+1)(2n+1)}{24}\right)$, thus for the critical value W it becomes:

$$W = \frac{n(n+1)}{4} + 1.96\sqrt{\frac{n(n+1)(2n+1)}{24}} = 608$$

Given the inequality shown by 497 < 608, it is not possible to reject the null hypothesis and we have to accept that the mean peak discharge is 200 [m³/s].

3.6.2 Homogeneity Tests

Student's t Test and the Fisher-Snedecor Test

The test for the homogeneity of the mean is based on *Student's two-sample t-test* while the test for homogeneity of variance corresponds to the *Fisher-Snedecor* test. Again, the reader is invited to consult a classic book on statistics for further details. Both of these tests are parametric and assume normal distribution of the population.

P

Example

Due to the fact that the Vispa watershed had been affected by a human intervention in 1964, the peak discharge series is divided into two samples:

$x_1, x_2, \ldots, x_{42}$ (peak discharges from 1922 to 1963)

and

$y_1, y_2, \ldots, y_{33}$ (peak discharges from 1964 to 1996).

Given that for Student's t-test we assume that the variances are equal but unknown, it would therefore seem judicious to first carry out the Fisher-Snedecor test. In this case, we have:

$$\begin{cases} H_0 : \sigma_x = \sigma_y \\ H_1 : \sigma_x \neq \sigma_y \end{cases} \qquad (3.14)$$

The discriminant function[13] is written as:

$$F_{obs} = \frac{\dfrac{n_1 s_x^2}{n_1 - 1}}{\dfrac{n_2 s_y^2}{n_2 - 1}} = 1.31 \qquad (3.15)$$

[13] In practice, for the discriminant function we always put the larger of the two quantities $n_1 s_x^2/(n_1 - 1)$ and $n_2 s_y^2/(n_2 - 1)$ at the numerator, thus the critical region is in the form $F > k$ with $k > 1$.

The critical value is $F_{n_1-1,\, n_2-1}(97.5\%) = 1.72$. Given that 1.31 is lower than 1.72, we cannot reject the null hypothesis stating that the variances are equal, and consequently we can apply Student's t-test for the two samples.

Knowing the impacts of a dam construction, we can expect to see a significant reduction in discharges in the second sample, which guides us in formulating a unilateral alternative hypothesis for testing the homogeneity of the samples based on the mean value.

This gives us:

$$\begin{cases} H_0 : \mu_x = \mu_y \\ H_1 : \mu_y < \mu_x \end{cases} \tag{3.16}$$

and

$$t_{obs} = \frac{\bar{x} - \bar{y}}{\sqrt{\dfrac{s_p^2(n_1 + n_2)}{n_1 n_2}}} = -5.9 \text{ with } s_p^2 = \frac{\left[(n_1 - 1)\, s_x^2 + (n_2 - 1)\, s_y^2\right]}{n_1 + n_2 - 2}$$

The critical value is given by $q_{t_{n_1+n_2-2}}(95\%) = 1.66$. Thus, we reject the null hypothesis because 1.66 < 5.90 and, as we had expected, the average peak discharge decreased significantly after the construction of the dam.

To test for the homogeneity of data from two populations we may apply two equivalent statistics – one from Mann-Whitney and the other from Wilcoxon (Helsel and Hirsch, 2002, page 118), as well as the median test, which is nonparametric.

Wilcoxon Homogeneity Test

Example

NF

Since we expect to see a significant reduction in discharge after 1964, we set the following hypotheses:

$$\begin{cases} H_0 : \mu_x = \mu_y \\ H_1 : \mu_y < \mu_x \end{cases} \tag{3.17}$$

Discharges indicated in bold (Table 3.5) belong to the second (post-dam) series.

Table 3.5 Application of the Wilcoxon homogeneity test.

discharge	**49**	55	63	76	...	**125**	125
rank	1	2	3	4	...	26.5	26.5
discharge	135	138	...	**270**	**285**	330	375
rank	28	29	...	72	73	74	75

Wilcoxon's *W* statistic is the sum of the ranks of the first sample. This means we have:

$$W_x = 28 + 29 + \dots + 72 + 75 = 2174,$$

$$W_y = 1 + 2 + \dots + 26.5 + 26.5 + \dots + 73 + 74 = 676$$

For $n_1, n_2 > 10$, we use the following normal approximation:

$$W_x \sim N\left(\frac{n_1(n_1 + n_2 + 1)}{2}, \frac{n_1 n_2(n_1 + n_2 + 1)}{12}\right) = N(1596, 8778)$$

The critical value is 1750. Since $W_x > 1750$, we reject the null hypothesis, which is consistent with our expectation.

Median Test

NP

Let us consider a sample of size n with x_i values (a time series, for example) with a median denoted $\tilde{x}$ (an alternative test involves using the average $\bar{x}$). Each observation x_i is assigned a "+" sign if it is greater than the median, and a "–" sign if it is smaller. Any group with "+" values is called a positive sequence and a group with "–" values is called a negative sequence. To apply the median test we have to determine N_s which is the total number of positive or negative sequences as well as T_s, the size of the longest of these sequences, which follows a binomial distribution. This leads to the following distribution:

$$N_s \sim N\left(\frac{n+1}{2}, \frac{n-1}{4}\right) \tag{3.18}$$

For a significance level between 91 and 95%, the verification conditions of the test are the following:

$$\frac{1}{2}(n + 1 - 1.96\sqrt{n-1}) < N_s < \frac{1}{2}(n + 1 + 1.96\sqrt{n-1}) \text{ and} \tag{3.19}$$

$$T_s < 3.3(\log_{10} n + 1) \tag{3.20}$$

If conditions (3.19) and (3.20) are verified, we cannot reject the null hypothesis stating the homogeneity of the series.

Example

We would like to test the homogeneity of the peak discharge series of the Vispa River for the total observation period, the median discharge being 150.

Table 3.6 Application of the median test.

discharge	240	171	186	158	...	145	155
sign	+	+	+	+	...	–	+
discharge	230	270		330	55	63	49
sign	+	+		+	–	–	–

We have $N_s = 22$ and $T_s = 9$ (see Table 3.6). Since $N_s < \frac{1}{2}(n + 1 - 1.96\sqrt{n-1}) = 29.5$, we have to reject the null hypothesis.

3.6.3 Goodness-of-fit Tests

Chi-square test

The parametric test used to test goodness of fit, based on the comparison of the theoretical and empirical frequencies, is the *chi-square test* (Helsel and Hirsch, 2002)[14].

Example

We would like to know if the series of peak discharges discussed above is distributed according to a normal distribution. Let Z be the random variable modeling the discharges. This gives us the following null and alternative hypotheses:

$$\begin{cases} H_0 : Z \sim N(\mu_z, \sigma_z^2) \\ H_1 : Z \text{ is not distributed according to a normal distribution} \end{cases} \tag{3.21}$$

The two parameters of the normal distribution are estimated by the sample mean and the sample variance, respectively. This gives us:

$$\hat{\mu}_z = \bar{z} = 158.14 \text{ and } \hat{\sigma}_z^2 = s_z^2 = 3967.4$$

Then we proceed to split the observations into 12 classes – this is an arbitrary choice – and after computing the observed and theoretical frequencies, we finally obtain $\chi^2_{obs} = 14.12$. This value is compared with the quantile of the chi-square distribution with $12 - 1 - 2 = 9$ degrees of freedom, that is to say $\chi^2_9\ (95\%) = 16.92$. Because $(14.12 < 16.92)$ it is not possible to reject the null hypothesis and we must conclude that the discharge follows a normal distribution. It should be mentioned that this result is not surprising given the fact that the chi-square test has a small power. It would have been wiser to select a goodness-of-fit test adapted to a normal distribution, such as Lilliefors or Shapiro-Wilks.

3.6.4 Autocorrelation Tests

It should be mentioned that the simplest and quickest method for evaluating serial independence, often explained in statistical hydrology books, consists of computing the lag-one autocorrelation coefficient of a series, and then applying one of the "traditional" parametric or nonparametric tests suggested for the "standard" correlation coefficient r_{xy}, that is to say:

---

[14] Also see Section 6.2.

- Fisher's test of the nullity of Pearson's correlation coefficient.
- Spearman's test of the nullity of the rank correlation coefficient.
- Kendall's test of the nullity of Kendall's correlation coefficient (Kendall's tau).

These three tests show a valid *nominal* behavior (see boxed text) when they are applied to a "traditional" bivariate series, but prove to be inapplicable in the case of the autocorrelation coefficient test, as shown by the results of a specific benchmark study (Meylan and Musy, 1999).

> The word ***nominal*** is used here only to simplify the terminology, and so needs a few clarifications: A test is said to be used in a *nominal framework* when it is applied *in the conditions that are at the basis of its development*. This implies that we respect both the reason for which it was conceived [here, as a rule, in a "classical" test of autocorrelation, for a comparison between the two terms of a bivariate series (x, y)] and as well, that the hypotheses on which it depends are respected [usually we use the assumption of a normal distribution for the series to be processed].
>
> This means that before running a test, we first verify that it behaves correctly within the setting of its "original" or **nominal** area of usage.

Below we discuss Kendall's test for the nullity of a correlation coefficient. The reader is invited to refer to some classical books on statistics, such as Saporta (1990), for details on the other tests.

Kendall's Test

The nonparametric test by Kendall (1938) aims to detect a monotonic dependence between two variables by means of a sample of n pairs (x_i, y_i). Computation of Kendall's statistic consists of counting a concordance (**+1**) if, for each possible pair (y_i, y_j), the corresponding pair (x_i, x_j) respects the same rank order. Otherwise, a discordance (**–1**) is counted. Kendall's statistic K is the algebraic sum of these "scores" or in other words, the difference between the number C of concordances and the number D of discordances.

NP

C, D and K are linked by the following equation (see for example Lebart *et al.*, 1982, page 153):

$$C = \frac{1}{2}\left(\frac{n(n-1)}{2} + K\right);\ D = \frac{1}{2}\left(\frac{n(n-1)}{2} - K\right) \tag{3.22}$$

The computation of the K statistic is made much easier if all the x_i values are first classified in ascending order (Lebart *et al.*, 1982, page 152).

In addition, the K statistic makes it possible to define a correlation coefficient, which is known as *Kendall's tau*:

$$\tau = \frac{2K}{n(n-1)} \tag{3.23}$$

The distribution of the correlation coefficient can be obtained by enumerating the $n!$ permutations of rank orders. This distribution has been tabulated, for example in Lebart *et al.* (1982, page 500) or in Siegel (1988).

For a sample size n larger than 10, it becomes possible to use an approximation through a normal distribution with parameters:

$$\mathrm{E}[\tau] = 0 \; ; \; \mathrm{Var}[\tau] = \frac{2(2n+5)}{9n(n-1)} \tag{3.24}$$

Kendall's test was applied to a lag-one autocorrelation coefficient. To do this, 10,000 independent and identically distributed samples were generated with the help of a normal distribution. The sample sizes varied from 5 to 500. Figure 3.14 shows the proportion of rejections of the null hypothesis for the unilateral test ($H_1 : \rho > 0$) for various thresholds (1, 2, 5 and 10%).

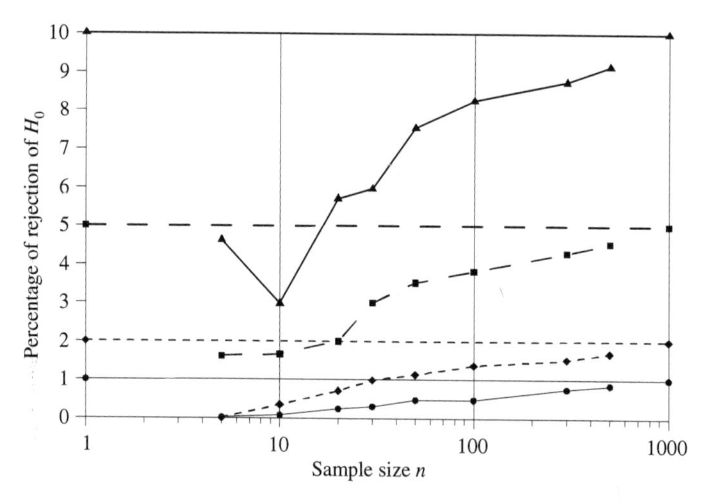

Fig. 3.14 The performance of Kendall's test for a lag-one autocorrelation coefficient. Proportion of rejection of nullity for thresholds: 1 (solid line – dot), 2 (dashed line – diamond), 5 (dashed line – square) and 10% (solid line – triangle).

As the observations are independent, if the test were valid, the rejection percentage should correspond to a type I probability of error (and thus to the threshold). Figure 3.13 shows that the test is not well adapted for an autocorrelation coefficient.

Anderson's Test

P

Anderson (1942) studied the distribution of the autocorrelation coefficient for a normal parent population. This leads to a parametric test. In this case, the autocorrelation coefficient is computed for n pairs of values (x_1,x_2), (x_2,x_3), ..., (x_{n-1},x_n), and (x_n,x_1).

For a "reasonable" sample size n (Anderson set a limit of 75 values!), the autocorrelation coefficient is distributed according to a normal distribution with mean and variance values:

$$\mathrm{E}[R] \approx -\frac{1}{n-1};\ \mathrm{Var}[R] \approx \frac{n-2}{(n-1)^2} \tag{3.25}$$

For smaller sample sizes, the distribution is more complex. Consequently, we provide tables of the critical values for the correlation coefficient (Table 3.7).

Again, a test of nominal behavior was carried out by generating 10,000 samples from a normal parent population. Sample sizes varied from 5 to 500. Because the sample sizes were small, the critical value from Table 3.7 was used. Figure 3.15 shows that in this case the test performs very well.

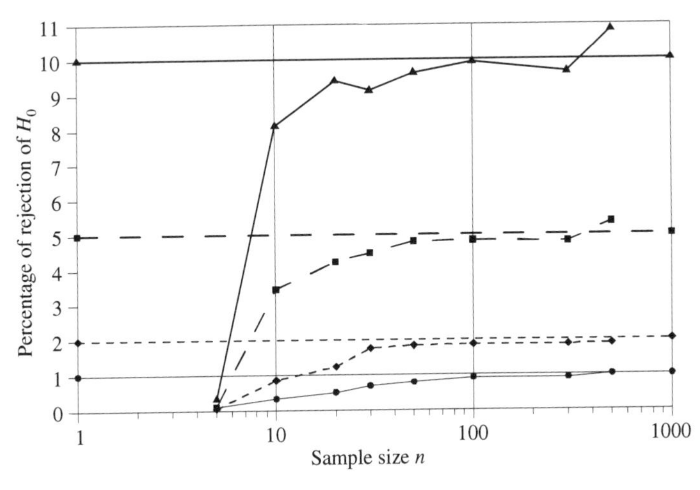

Fig. 3.15 Nominal performance of Anderson's test. Proportion of rejection of the null hypothesis for thresholds of 1% (solid – dot) and 5% (dashed – square).

Table 3.7 Table of Anderson's critical values for the autocorrelation coefficient for unilateral tests.

n	20%	10%	5%	2%	1%
5	0.0854	0.1972	0.2531	0.2867	0.2978
6	0.1085	0.2534	0.3446	0.4157	0.4469
7	0.1156	0.2644	0.3696	0.4629	0.5099
8	0.1225	0.2641	0.3713	0.4744	0.5307
9	0.1310	0.2622	0.3661	0.4714	0.5322
10	0.1361	0.2611	0.3597	0.4632	0.5253
11	0.1378	0.2592	0.3538	0.4537	0.5153
12	0.1382	0.2558	0.3478	0.4445	0.5047
13	0.1381	0.2517	0.3413	0.4357	0.4944
14	0.1375	0.2476	0.3346	0.4269	0.4846
15	0.1366	0.2434	0.3281	0.4183	0.4749
16	0.1355	0.2393	0.3218	0.4100	0.4656
17	0.1342	0.2353	0.3158	0.4021	0.4566
18	0.1329	0.2314	0.3100	0.3944	0.4480
19	0.1314	0.2276	0.3044	0.3872	0.4398
20	0.1300	0.2239	0.2991	0.3803	0.4320
21	0.1285	0.2204	0.2941	0.3737	0.4245
22	0.1271	0.2170	0.2892	0.3674	0.4174
23	0.1257	0.2138	0.2846	0.3614	0.4106
24	0.1242	0.2107	0.2801	0.3556	0.4041
25	0.1229	0.2077	0.2759	0.3502	0.3979
26	0.1215	0.2048	0.2718	0.3449	0.3919
27	0.1202	0.2020	0.2679	0.3399	0.3862
28	0.1189	0.1993	0.2642	0.3351	0.3808
29	0.1176	0.1967	0.2606	0.3304	0.3755
30	0.1164	0.1943	0.2572	0.3260	0.3705
35	0.1107	0.1831	0.2418	0.3062	0.3480
40	0.1057	0.1736	0.2288	0.2896	0.3291
45	0.1013	0.1655	0.2178	0.2754	0.3131
50	0.0974	0.1585	0.2082	0.2632	0.2991
55	0.0939	0.1522	0.1998	0.2524	0.2869
60	0.0908	0.1467	0.1923	0.2429	0.2761
65	0.0880	0.1417	0.1857	0.2344	0.2665
70	0.0854	0.1373	0.1797	0.2268	0.2578
75	0.0830	0.1332	0.1742	0.2198	0.2499
80	0.0809	0.1295	0.1692	0.2135	0.2427
85	0.0789	0.1260	0.1647	0.2077	0.2361
90	0.0770	0.1229	0.1605	0.2023	0.2300
95	0.0753	0.1199	0.1566	0.1974	0.2243
100	0.0737	0.1172	0.1529	0.1928	0.2191

It should be noted that the Anderson test, which is parametric, is very *robust*, comparable in this regard to the nonparametric test from Wald-Wolfowitz, which is described below.

Example

Let us test the nullity of the lag-one autocorrelation coefficient in our first discharge series: the discharge of the Vispa River from 1922 to 1963. The hypotheses for this test are therefore:

$$\begin{cases} H_0 : \rho_1 = 0 \\ H_1 : \rho_1 > 0 \end{cases} \tag{3.26}$$

If we estimate the lag-one autocorrelation coefficient in our data set (1922 to 1963) we obtain a value close to 0, or in fact 0.002. As, according to Table 3.7, the critical value is about 0.22, we cannot reject the null hypothesis. It should be noted that this result was expected because we are dealing with an annual series where *a priori* the persistence effect is zero.

Wald-Wolfowitz Test

Based on Anderson's work, Wald and Wolfowitz (1943) developed a nonparametric test of the autocorrelation coefficient. This statistic is calculated as follows:

NP

$$R = \sum_{i=1}^{n-1} x_i x_{i+1} + x_n x_1 \tag{3.27}$$

For a "sufficiently" large n, this statistic follows a normal distribution with mean and variance:

$$\mathrm{E}[R] = \frac{S_1^2 - S_2}{n-1} \tag{3.28}$$

$$\mathrm{Var}[R] = \frac{S_2^2 - S_4}{n-1} + \frac{S_1^4 - 4S_1^2 S_2 + 4S_1 S_3 + S_2^2 - 2S_4}{(n-1)(n-2)} - \{\mathrm{E}[R]\}^2$$

with

$$S_k = \sum_{i=1}^{n} x_i^k \tag{3.29}$$

It should be remembered that nonparametric tests do not depend upon precise parameters or assumptions concerning the subjacent distribution.

3.6.5 Tests for Trend

Kendall's Test for Trend

This is a nonparametric test for trend. Here, we will use Sneyers' formulation (1975). For each observation x_i we determine the number n_i of observations x_j preceding it ($j < i$) and which are inferior to it ($x_j < x_i$). This means the statistic is simply the sum of n_i:

$$t = \sum_{i=1}^{n} n_i \tag{3.30}$$

which for a large enough n follows a normal distribution with parameters:

$$\mathrm{E}[t] = \frac{n(n-1)}{4} \tag{3.31}$$

and

$$\mathrm{Var}[t] = \frac{n(n-1)(2n+5)}{72} \tag{3.32}$$

Sneyers (1975) put forward a clever version of this test which involves visualizing the evolution of the centred reduced statistic $z(t_i)$ in relation to the temporal abscissa i as well as $z'(t_i)$ obtained through a retrograde calculation of n to 1. It is interpreted as follows:

- In the absence of any trend we obtain two curves that are tangled.
- If there is a marked trend, the intersection of the two curves allows us to pinpoint the start date of the rupture of the "non-nullity trend."

The most interesting characteristic of this test is certainly its graphical representation.

Example

Once again using the peak discharge series of the Vispa River at Visp, we get Figure 3.16, where we can locate the anomaly that took place around 1963:

The test statistic results in $|z| = 5.86$. Therefore in the absence of a trend, we must reject the null hypothesis.

Bois "Monostation" Test

The Bois test (cumulative residuals) is described in its original form in section 3.7.2.

This test can be adapted to advantage for a "monostation" format as follows: The test aims to compare the checked station to a fictitious

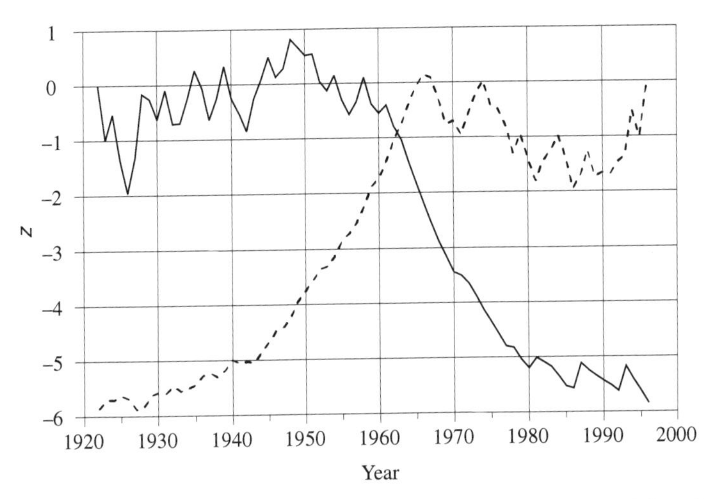

Fig. 3.16 Kendall's progressive test of trend; solid line shows z, dashed line shows z'.

station built with the station mean. This means looking at the following cumulative residuals relation:

$$\varepsilon_i = x_i - \bar{x} \tag{3.33}$$

to be compared to equation (3.36).

Example

With the same data as above (the Vispa River at Visp) we obtain the cumulative residuals represented in Figure 3.17, which illustrates the trend in the series.

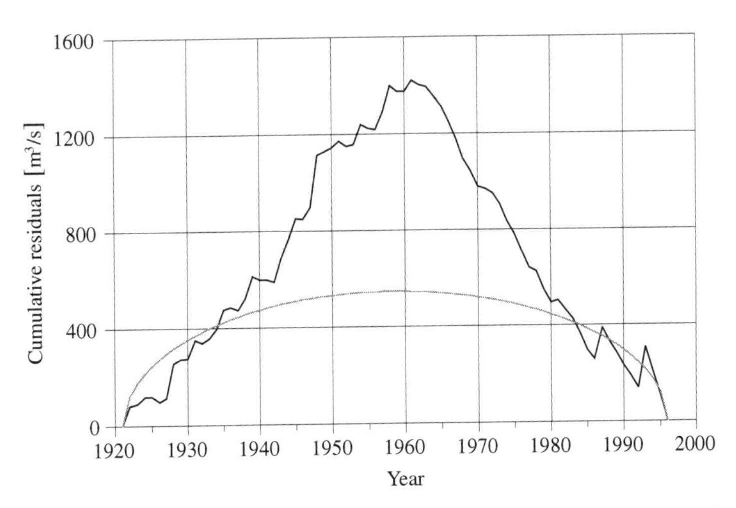

Fig. 3.17 Annual peak discharge of the Vispa River at Visp. Bois *monostation* test. Confidence ellipse at 95%.

3.7 ONE PARAMETER – REGIONAL SCALE

These methods are employed to check, for example, the rainfall data from several stations in a watershed.

The two tests discussed in this section are mainly used in hydrology. Their specific purpose is to compare one or several samples, obtained from neighboring stations, in order to detect a possible lack of homogeneity (the most common reason is a change at one of the stations, for example the displacement of the raingauge).

3.7.1 Double Mass Method[15]

The principle of this method is to check the proportionality of the values measured at two stations. One of the stations (station X) is the base station or reference station, and considered to be correct. The other station (Y) is the station to be checked. A smoothing effect is achieved by comparing, for a selected time step (year, season, month, decade), the cumulative value rather than the actual measured values. The method is conceptually very simple, since it involves simply drawing a graph of the following quantities:

$$X(t) = \sum_{i=1}^{t} x(i) \text{ and } Y(t) = \sum_{i=1}^{t} y(i) \tag{3.34}$$

Example

We want to graphically test the homogeneity of the peak discharge data for the Vispa River at Visp.[16] To do this, we will use for our reference station the discharge of the Rhone at Brig (after ascertaining that this series is homogenous). Figure 3.18 shows how the double mass method applies in such a situation. For the station we are checking, the graph shows a clear disruption in the slope for the year 1964. This method is able to detect an anomaly (the construction of the dam) but does not aim to correct it.

The advantage of the double mass method is that it is simple, fast, and well-known. On the other hand, however, it is not always easy to interpret the graphs, and even more important, the method does not provide any scale of probability for the defects detected, so it does not serve as a test in the statistical sense. Lastly, although this method can detect errors, it cannot correct them, at least not directly. However if this method reveals a critical situation, a correction can be made following a more in-depth analysis.

---

[15] Also called by statisticians "CUSUM method."

[16] The reference station can be replaced if appropriate with a fictitious station representing an ensemble of "reliable" stations in the area (see section 3.7.3).

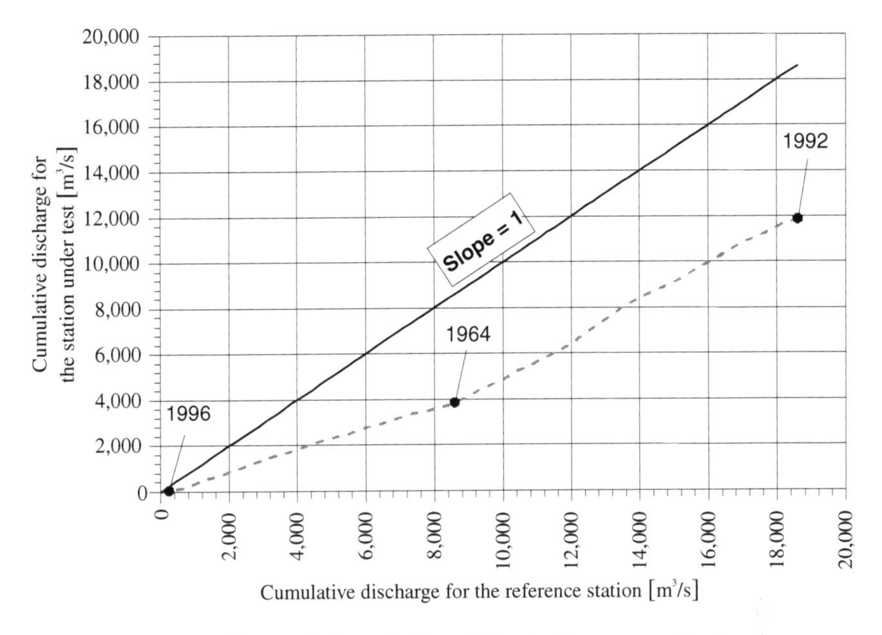

Fig. 3.18 Example of a practical application of the double mass method.

3.7.2 Cumulative Residual Method

NP

The cumulative residual method, credited to Philippe Bois of the Ecole nationale supérieure d'hydraulique de Grenoble (Bois, 1986), is an extension of the idea of the double mass method, with the addition of statistical content allowing the application of a true homogeneity test – an important advance.

Again for a double series of values x_i (reference series) and y_i (series to be checked), the basic idea is to study, not the values x_i and y_i (or $\sum x_i$ and $\sum y_i$) directly, but the cumulative of the residuals ε_i resulting from a linear regression between y and x:

$$y_i = a_0 + a_1 x_i + \varepsilon_i \tag{3.35}$$

$$\varepsilon_i = y_i - (a_0 + a_1 x_i) = y_i - \hat{y}_i \tag{3.36}$$

Figure 3.19 shows such a regression.

According to classical regression theory, the sum of the residuals is by definition null with a normal distribution and standard deviation:

$$\sigma_\varepsilon = \sigma_Y \sqrt{1 - r^2} \tag{3.37}$$

where r is the Pearson linear correlation coefficient between X and Y.

For a sample with n *counts*, the cumulative residual E_j is defined by:

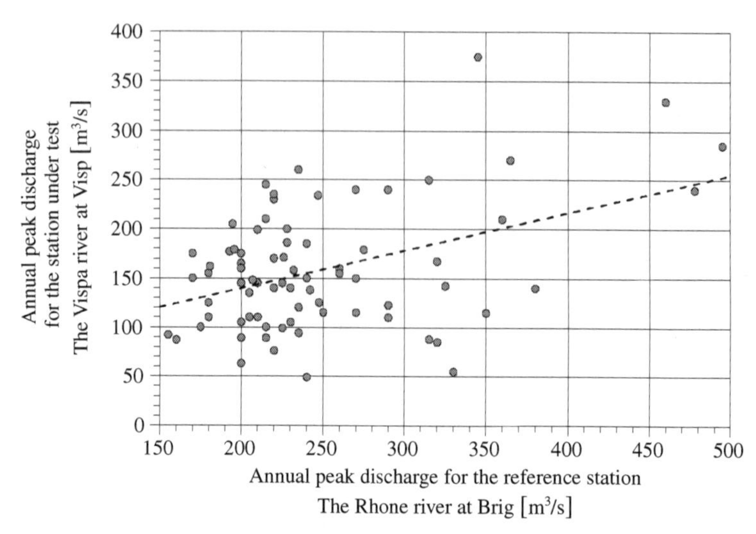

Fig. 3.19 Dispersion diagram of annual peak discharges of the Rhone River at Brig and the Vispa River at Visp (1922-1996).

$$E_0 = 0; \; E_j = \sum_{i=1}^{j} \varepsilon_i \;\; \forall j = 1, 2, \ldots, n \tag{3.38}$$

The graphic report of the cumulative residuals E_j (on the ordinate) in relation to the serial numbers j of the values (on the abscissa, $j = 0$ at n, with $E_0 = 0$) should, for a proven correlation between X and Y, give a line starting at 0, oscillating randomly around zero between $j = 0$ and $j = n$ and ending at 0 for $j = n$. The presence of a lack of homogeneity is shown by non-random deviations around the zero value.

Bois described and tested numerous types of nonhomogeneity. Among other things, he demonstrated that, for a selected level of confidence $1 - \alpha$, the graph of E_j as a function of j must be inside an ellipse on the large axis n and the ordinate:

$$\pm z_{1-\alpha/2}\sigma_\varepsilon\sqrt{\frac{j(n-j)(n-1)}{n^2}}, \; j = 0, \ldots, n \tag{3.39}$$

where $z_{1-\alpha/2}$ is the $\left(1 - \frac{\alpha}{2}\right)^{th}$ quantile of the standard normal distribution.

These developments provide a true test of homogeneity between two stations.

Figure 3.20 shows Bois' test applied to the data from the two measuring stations in our earlier examples. We can see a change in the behavior around the year 1964.

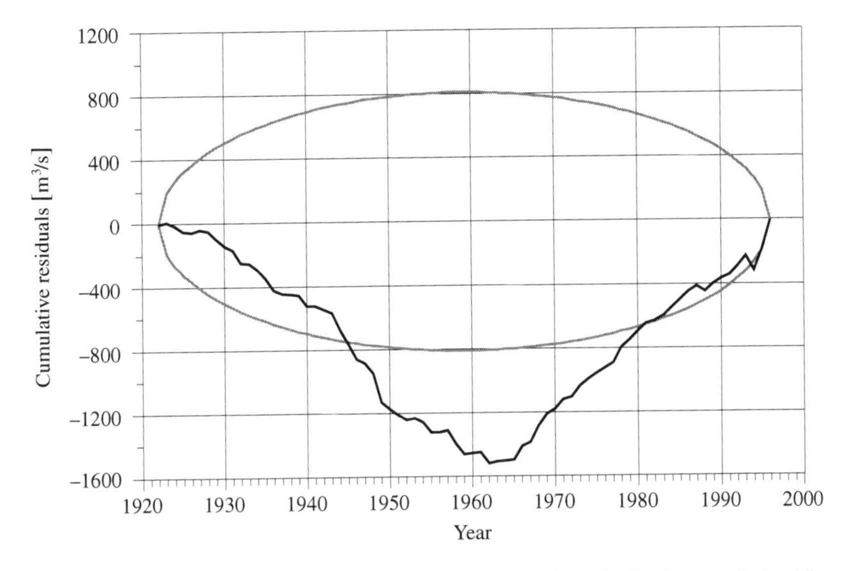

Fig. 3.20 Result of Bois' test as applied to the annual peak discharge of the Vispa River at Visp using the discharge of the Rhone at Brig as the reference series. Confidence ellipse of 95%.

3.7.3 Fictitious Regional Station

The two techniques described briefly below are for the purpose of explaining a *fictitious "regional" station* that is characteristic of a group of stations. This fictitious station, representative of the entire area, can then be used as the *reference station* to check the individual stations using the double mass method or the Bois ellipse method.

These techniques are both based on the principle of *pseudo-proportionality* but they use different estimation techniques: essentially this involves taking into account possible errors in the data by a fairly complex iterative process which is not described here. The "regional vector" solution is just one aspect of these techniques, the other being the homogenization of data.

Regional Vector Method

This technique, developed by G. Hiez (1977), consists of determining a *regional vector* $\vec{l}_i$ of representative indices of annual rainfall for a region while quantifying the relative contribution $\vec{c}_i$ of each station. Thus the regional vector represents, in compact form, the most probable rainfall trend for the region in consideration.

The model used is:

$$\mathbf{A} = \mathbf{B} + \mathbf{E} \tag{3.40}$$

where **A** is the matrix of observations with dimension n (number of years) times m (number of stations), **B** is the "theoretical" data matrix and **E** is the error matrix (also called the residual matrix).

Matrix **B** is obtained by applying the principle of *pseudo*-proportionality:

$$\mathrm{B} = \vec{c}\vec{l} \tag{3.41}$$

where $\vec{c}$ is the column vector with n elements *of annual indices* and $\vec{l}$ is the row vector of m *station coefficient* elements.

The calculation technique consists of defining the $\vec{l}$ and $\vec{c}$ elements so as to minimize the residual matrix using an algorithm called the *l-c process* (row-column).

Annual Rainfall Indices Vector

Brunet-Moret (1979) defined a vector $\vec{z} = (z_1, \cdots, z_j, \cdots, z_n)$ of annual precipitation indices using an equation of *pseudo-proportionality*:

$$\frac{P_{ij}}{\overline{P}_i} = z_j + e_{ij} \tag{3.42}$$

where P_{ij} is the precipitation module for year j at station i, $\overline{P}_i$ is the normal for the station i, z_j denotes the annual precipitation index for year j and e_{ij} is a random residual.

The principle of the method consists of simultaneously determining the values of the regional index z_j for each year j and the values of the normals $\overline{P}_i$ for each station i while minimizing the sum of squared deviations e_{ij}, based on the following hypotheses:

- the expectation of z_j is one;
- the expectation of the residuals e_{ij} is null;
- the variances for each station i (e_{ij}) are all equal.

CHAPTER 4

Selecting a Model

The choice of a model in a frequency analysis is without doubt the most critical step, and the one that introduces the largest uncertainties. This chapter explains why there is no "simple recipe" for this selection, looks at some common practices, and explores some theoretical perspectives.

4.1 THE GAMBLE OF FREQUENCY ANALYSIS

4.1.1 Natural Processes and Theories of Probability

The use of a *frequency model* $F(x, \alpha, \beta, \ldots)$, in order to *describe*, in operational and practical form, the hydrological phenomenon (or process) under study is a *fundamental decision*: this decision determines the framework of the study and the tool to be used: the probability model (Matheron, 1989, page 4).

Matheron also says: "*In fact there is not, nor can there be, any such thing as probability in itself.* There are only probabilistic models. *In other words, randomness is in no way a uniquely defined property, or even definable property of the phenomenon itself. It is only a characteristic of the model or models we choose to describe it, interpret it, and solve this or that problem we have raised about it.*" Adopting a frequency model in order to study and describe hydrological phenomena is thus a *decision*, a *choice*: the motivations for this choice, and its consequences, are the subject of the following paragraphs.

4.1.2 Some Reasons for Choosing a Probability Model

The Complexity of Phenomena

In the field of hydrology, the phenomena involved are so complex that even if a particular process might be considered as completely deterministic, it always has a random nature. Essentially, it is often impossible to determine, or even identify, all the parameters that affect the process, or to know exactly which physical laws regulate it. Yevjevitch (1972) even claimed that there is no hydrological phenomenon that can be considered as purely deterministic.

This means, then, that the complexity of the phenomena we encounter in hydrology tends to push us towards statistical types of analyses in order to understand them.

The Statistical Nature of the Expected Answers

The information that an engineer wants to introduce in a particular study is very often of a *statistical* nature:

- Analyzing risk $R = pc$ (see Section 2.3) requires determining p, the annual occurrence probability of the event under study;
- The same annual probability $p = 1/T$ makes it possible to compute various scenarios for more complex events (see Section 2.2);
- It is very often necessary to be able to estimate the uncertainties affecting a result; this computation requires the formulation of a standard error (see Chapters 7 and 8) and certain composition rules for errors.

This means that the choice of a frequency model, which *de facto* includes a probability component (p, $1/T$, $x(F)$ or $F(x)$), is also in part motivated by the nature of the answers being sought.

Interpolation and Extrapolation

In the field of frequency analysis, *interpolation* consists of estimating the events for $T << n$, where n is the number of years of observations available. This means that the estimation of an event with a return period of 10 years based on 50 to 100 years of data is an interpolation.

The opposite procedure – and the one that is most commonly needed in practice – is extrapolation, which consists of estimating the events for $T > n$. Figure 4.1 illustrates the differences between interpolation and extrapolation. The estimation of an event with a return period of 100 years using extrapolation based on about 30 years of observations, as it is currently done, is invariably accompanied by some major conceptual problems.

According to Klemeš (1986): "*Extrapolation of flood frequency curves to obtain estimates of the customary 100-, 500-, 1000-year, etc., flood has neither a sound empirical basis nor a theoretical one. From an empirical point of view, a 2- or 5-year flood may be a sound concept on the assumption that 50 years or so of data are available, the historic record does not look conspicuously different from a random series, and physical conditions during that period are known to have been approximately stationary.*"

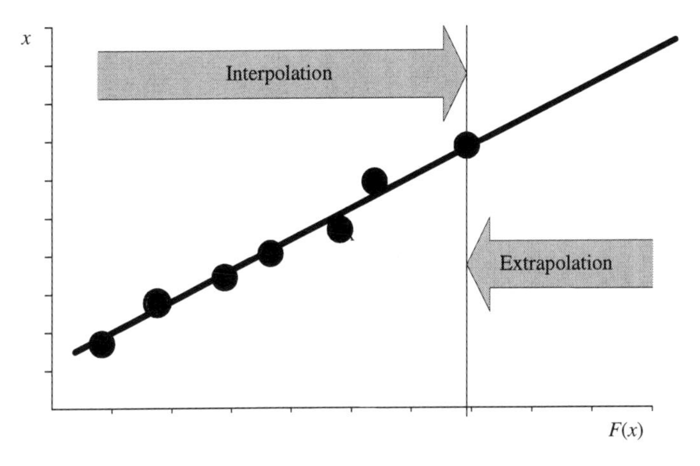

Fig. 4.1 Interpolation and extrapolation in frequency analysis.

Along similar lines, Wallis (1988) commented: "… *unless the distribution is known, at-site estimates of x(F) (i.e. estimates made from a single sample of data) should neither be calculated nor used. A possible exception to this rule might be in those situations where T<<n.*"

Wallis advocates a different approach: the so-called regional approach will be addressed in Chapter 9.

4.1.3 The Steps in Modeling

Matheron clearly described the different steps needed in the modeling process (1989, page 52 and following), and his explanation is cited in part below. The resulting modeling principle is illustrated in Figure 4.2.

Choice of the Constituent Model

This is first and foremost an epistemological choice: we decide to make use of a frequency model in order to reproduce a given phenomenon. It is, then, a *decision*, and not a controllable experimental hypothesis: the *constituent decision*.

Choice of the Type of Model

Matheron adds: "*Contrary to the preceding choice this second choice follows from a physical hypothesis which can be objectively tested. It can, therefore, be supported or rejected by the experimental data either through statistical tests … or through some other method, including the judgment of the practitioner.*"

[...] "*We cannot overstress the capital importance of this step. For it is essentially here that we incorporate in the model hypotheses that have an objective meaning and that carry with them positive information which is not*

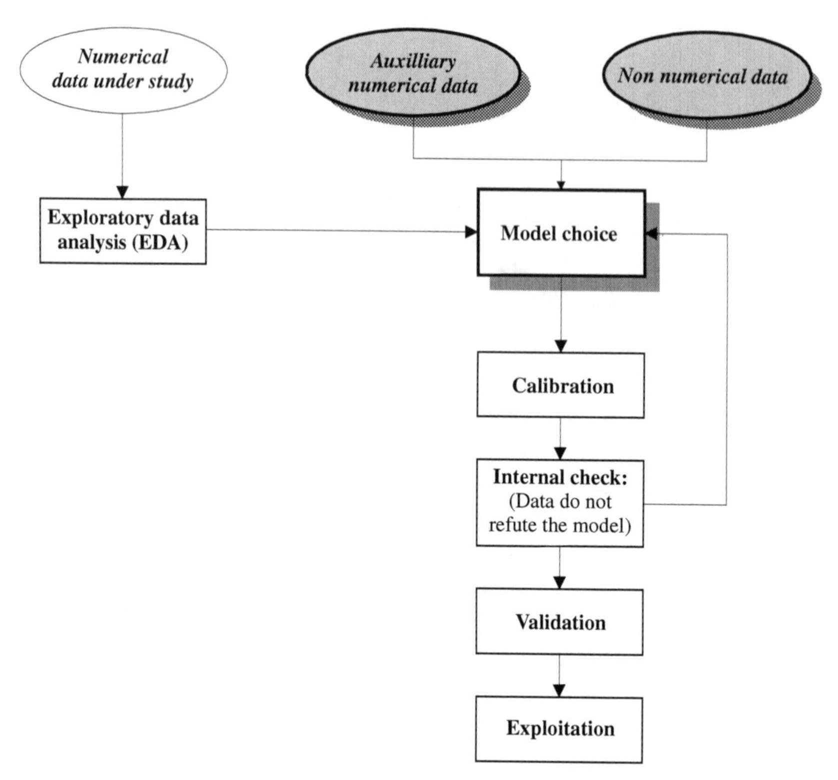

Fig. 4.2 Modeling principle and steps.

contained in the raw data. It is only because of this positive contribution that we can (apparently) extract from the data more than they really contain. The counterpart of this small-scale miracle is that our model is now vulnerable, and that our predictions can now be contradicted by experiment if the hypotheses on which we base them are not objectively valid."

Calibration

The final step in the modeling process is the calibration of the model, or in other words, the estimation of the model parameters. One approach (for example the maximum likelihood method) will be better adapted for a given distribution than another method.

However it should be noted that the validity of frequency analysis depends much more on the suitability of the model adopted than on the numerical estimation technique used.

Consequences: The Gamble of Frequency Hydrology

In essence, then, frequency analysis depends on the *choice* of a type of model, a choice that in practice is difficult to validate. This method implies among other things the assumption that the behavior of the process is similar in areas where it is not known to where it has been observed. From the point of view of frequency hydrology, this assumption includes two aspects: the persistence of the behavior (stationarity of the process) but also, and above all, the problem of the behavior of the *tail of the distribution* (which is to say large values of $F(x)$), supposedly deductible from the behavior at *average* frequencies, which are the only ones usually available for measurement. With this hypothesis, frequency analysis seems to be a real gamble with multiple facets.

The second important element that results from the various aspects mentioned above relates to the fundamental role played by *non-numerical information*: experience from similar cases, expert knowledge, analogies to other phenomena, historical data, etc. All such information must be employed in support of the choice of model.

4.1.4 Can the Soundness of the Choice be Verified?

There is no denying that once a model has been chosen, it tends to follow its own course, often without any direct connection to the reality that it is supposed to be describing.

Matheron (1989, page 23) expressed the difference between the model and reality in these terms: "*The model is never identical to reality. There are innumerable aspects of reality which will always elude it, and on the other hand the model will always contain parasitic concepts, which have no counterpart whatsoever in reality.*"

Given the practical range of statistical tests, all that a goodness-of-fit test, for example, can do is allow us to conclude that the available data makes it possible or not to reject the hypothesis that the model is suitable for describing the observed values: such a test can never allow us to conclude that *another* particular model should be adopted. Even further, nothing can tell us whether the model is suitable for describing parts of the phenomenon that were not observed. However, whenever possible, we should still make an effort to validate the calibrated model, as discussed later.

At first glance, it would appear that verifying the suitability of the choice of model has been ignored in the field of frequency hydrology. Linsley (1986), for example, stated: "*Since they [practitioners] are estimating a relatively uncommon flood (at least a 10-year event, often a 100-year event), the probability that they will ever know how the estimate comes out is low. The probability that anyone will ever point a finger and say "you were wrong" is equally remote. If the estimate is exceeded, it is "obvious" that the new flood is more than the 10- or 100-year event, as the case may be. If the estimate is not exceeded, then there is no reason to think about it.*"

However, it is a best practice to attempt to validate using the proper procedure for any well-conducted modeling exercise. To that end, the available data are, ideally, divided into two subsamples:

- the first subsample is used to *calibrate* the model;
- the second set of data, which in theory is distinct from the first one, is used to *validate* the calibrated model.

However, this technique clearly reduces the useful sample size, which is often limited to begin with. In this situation we can apply resampling techniques (such as the "bootstrap" or "jackknife." The bootstrap technique will be discussed briefly in Chapter 7).

4.2 POSSIBLE APPROACHES TO CHOOSING THE TYPE OF MODEL

As we have just seen, the validity of the results of a frequency analysis depends very much on the *choice* of frequency model (and more particularly the *type*). Various approaches can be taken to facilitate this choice, but unfortunately there is no universal and infallible method.

4.2.1 Asymptotic Behavior

From the standpoint of asymptotic theory, the distributions of the maxima and of the values exceeding a given threshold are well known. We will briefly summarize them below.

For a sample $X_1, \ldots, X_n$ issuing from a usual distribution F it is possible to reach the distribution of the maxima of a sample by means of the generalized extreme-value distribution (GEV) defined by:

$$\Pr\{M_n = \max(X_1, \ldots, X_n) \le z\} \xrightarrow{n \to \infty} G(z),$$

where

$$G(z) = \exp\left[-\left\{1 + \gamma\left(\frac{z - \alpha}{\beta}\right)\right\}^{-1/\gamma}\right] \tag{4.1}$$

defined on $\left[z : \{1 + \gamma(z - \alpha)/\beta\} > 0\right]$ with $-\infty < \alpha < \infty$, $\sigma > 0$ and $-\infty < \gamma < \infty$.

This implies that we know an approximation of the distribution of the maximum of a sample of sufficient size coming from whatever distribution commonly used in hydrology. However, in practice we have no precise indication of the sample size required to make this approximation acceptable. A different asymptotic behavior is possible but would require building a model for which the precise hypotheses of the above theorem have not been verified (for example, see Embrechts *et al.*, 1997, page 158). Moreover because the maximum function being applied on a sample issued from a distribution of some kind, the usual hypothesis of independent and identically distributed random variables applies. For example, if we are dealing with annual maximum peak daily discharges, the basic sample is composed of daily values, which are not independent, (but autocorrelated up to a certain lag) and not homogenous (the series is made up of large and smaller discharges that do not issue from the same population because they do not necessarily result from the same hydrological phenomenon). However, it is possible to circumvent the hypothesis of independence (Coles, 2001, Chapter 5 and page 105).

The $G(z)$ model defined in Equation (4.1) is the generalized extreme-value distribution (GEV).

Note that the case where $\gamma > 0$ corresponds to the Fréchet distribution, where $\gamma < 0$ corresponds to the Weibull distribution and where $\gamma = 0$ corresponds to the Gumbel distribution. Each of these distributions induces an attraction domain, which is to say that a distribution belonging to one of the domains of attraction validates the asymptotic relation (4.1) corresponding to the particular sign of γ.

An important consequence of the convergence of the maximum distribution concerns the distribution of the *excesses* of a random variable X over a (sufficiently high) threshold u which converges conditionally towards a generalized Pareto distribution:

$$F_u(y) = \Pr(X - u \le y | X > u) \xrightarrow{n \to \infty} 1 - \left(1 + \frac{\gamma y}{\tilde{\beta}}\right)^{-1/\gamma} \tag{4.2}$$

defined on $\left\{y : y > 0 \text{ et } \left(1 + \gamma y/\tilde{\beta}\right) > 0\right\}$ with $\tilde{\beta} = \beta + \gamma(u - \alpha)$ (Pickands III, 1975).

It should be noted that the choice of the appropriate threshold is a statistical problem in itself and has not been fully resolved yet. This choice involves finding a compromise between the gain in information compared to the use of a single value per year (the maximum), and the independence of the series. Basically, in order to gain the maximum of information it would make sense to choose a reasonably low threshold, but the lower the threshold, the greater the risk of introducing dependence in our sample. Various methods can be found in the literature to guide our choice of threshold – the *Hill plot* (Hill, 1975), the *mean excess plot* (Davison and Smith, 1990), the bootstrap method based on mean squared error (Danielsson and de Vries, 1997; Danielsson *et al.*, 2001).

For some other approaches to determining the threshold, the reader is referred to Coles (2001) (p.78 to 80).

Note

The case $\gamma = 0$ is defined in the sense of a limit $\gamma \rightarrow 0$ and in this case, the maximum is given by an asymptotic Gumbel distribution, and the excesses by an exponential distribution.

It is important to point out here that in hydrology, the size of the available samples is relatively small. As a consequence, the hypothesis of asymptotic behavior can almost never be verified. This is why other types of distributions can be better adapted to small samples, and should also be considered.

A comparison of the behavior of different distributions for large values of $F(x)$ (or of x), or in other words, of the *tail* of the distribution, can be attempted. If, as is standard, the cumulative frequency $F(x)$ is transformed to a Gumbel variable u ($u = -\ln[-\ln\{F(x)\}]$), it becomes possible to distinguish four types of behaviors (Lebel and Boyer, 1989), as illustrated in Figure 4.3.

As interesting as this approach may be, it really only allows us to examine the *mathematical behavior* of the functions, whereas it is the *physical behavior* (or processes) which is unfortunately unknown, that is of primary interest. However it requires great prudence with logarithmic type distributions, which can provide excessively high values for extreme frequencies.

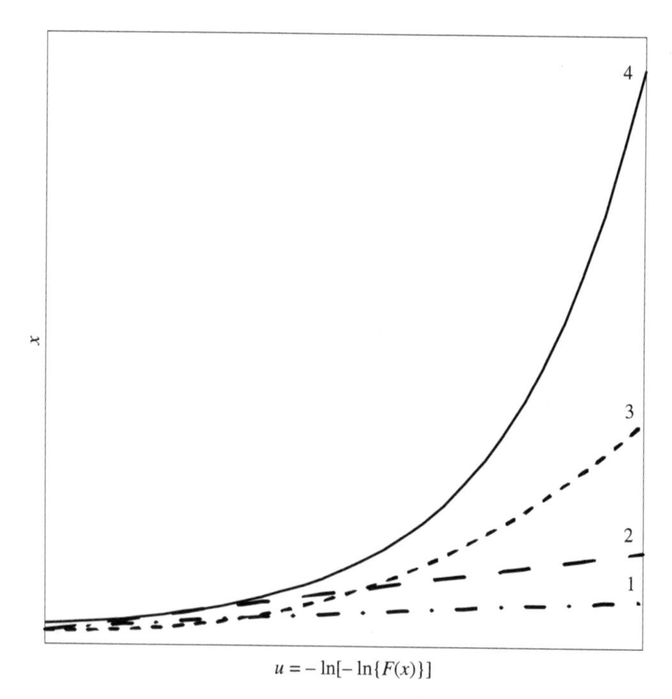

Fig. 4.3 Four types of asymptotic behavior:

1. $x \propto u^{\frac{1}{n}}$, with $n > 1$; for example: normal distribution, Goodrich distribution ($\beta < 1$)...;
2. $x \propto u$, distributions characterized by asymptotic exponential growth; for example the Gumbel, Pearson type III, and Goodrich ($\beta = 1$) distributions, the law of leaks;
3. $x \propto u^n$, with $n > 1$; for example the Goodrich distribution ($\beta > 1$);
4. $x \propto \exp(u^n)$, with $n > 0$, distributions of the logarithmic type; for example, the lognormal distribution (Galton), the Pearson type V, Fréchet and log gamma distributions.

4.2.2 Theoretical Considerations

Generalized Extreme Value Distribution (GEV)

As noted in the preceding paragraph, in an asymptotic context, the generalized extreme-value distribution (GEV) represents the distribution of the maximum of n random variables.

The Gumbel Distribution (GUM)

The Gumbel distribution is a particular case of the generalized extreme-value distribution where $\gamma = 0$. It is widely applied due to its ease of use.

The Generalized Pareto Distribution (GPD)

Again in an asymptotic context, the generalized Pareto distribution corresponds to the distribution of the excesses above a threshold of the generalized extreme-value distribution. It is used for modeling peak over a threshold (POT) series (*infra-annual* series, see Chapter 8).

The Exponential Distribution (EXP)

The exponential distribution is a particular case of the generalized Pareto distribution: It corresponds to the distribution of the excesses above a threshold of a Gumbel distribution, and therefore can also be used for modeling POT series.

The Normal Distribution (N)

In theory, the normal distribution (also called Gaussian or Gauss distribution) is explained by the central limit theorem, as the distribution of a random variable corresponding to the sum of a large number of random variables. In frequency hydrology, this distribution is quite often ruled out by the notorious lack of symmetry of the distributions. However it generally serves quite well for the study of annual averages of hydrometeorological variables.

The Lognormal Distribution (LN)

The *lognormal* distribution can be explained by arguing that the occurrence of a hydrological event results from the combined action of a large number of factors that multiply (Chow, 1954, for example). In such case, the central limit theorem leads to the conclusion that the random variable $X = X_1X_2...X_r$ follows a lognormal distribution (because the product of r variables yields to the sum of r logarithms). In the field of hydrology, this distribution is also known as the Galton, Gibrat or Galton-Gibrat distribution.

The Gamma Distribution (GAM)

The gamma distribution (and particularly the three-parameter gamma distribution) is very flexible (Bobée and Ashkar, 1991). It is usually applied for the frequency analysis of volumes corresponding to annual maximum floods. It should be noted that these volumes are not maximum values in the statistical sense (because peak floods that are lower but of longer duration can still produce larger volumes). It must be emphasized that the gamma distribution is a particular case of the Pearson type III distribution with a location parameter equal to zero. The terminologies Pearson type III distribution and three-parameter gamma distribution are equivalent and we will use the first one throughout this book.

4.2.3 Experience and Conventional Practice

The choice of the type of probability model is often based on local *habit,* which, over time, becomes actual *conventional practice,* or even *rules.* Frequently, the choice is made based on the experience of one or several authors, which eventually become *schools of thought.* The stance of the practitioner, who may follow one or another school of thought, is justifiable from the point of view that it allows him/her to benefit from a vast sum of experience, but it also carries some dangers. One of the manifestations of this can be what we call *contamination by citation,* which can result in the perpetuation of a choice, sometimes misguided, that is inadequate.

Another circumstance can likewise lead to the adoption of a particular model: In certain countries, their administrations have set *regulations, standards* or *directives* regarding the choice of methods to employ for frequency analysis. As an example, various U.S. states recommend the log-Pearson type III distribution for the analysis of annual maximum discharges.

4.2.4 Applying Goodness-of-fit Tests

Many authors apply *goodness-of-fit tests* (see Chapter 6) as tools for *choosing* the suitable frequency model. This way of proceeding is questionable for at least two reasons:

- First of all, in theory, a statistical test only makes it possible to reject, or not, the null hypothesis H_0. Statistical tests do not allow for a comparison between several null hypotheses (several frequency models) so as to choose the best one.
- Next, and more practically, the usual goodness-of-fit tests (for example the chi-square or Kolmogorov-Smirnov tests) suffer from a lack of power, meaning that they are not sufficiently able to reject the null hypothesis H_0 when the alternative hypothesis H_1 is actually true. It should be added that there exist some more powerful goodness-of-fit tests (such as Lilliefors, Shapiro-Wilks, Shapiro-Francia...) which are, however, only suitable for a normal distribution.

The GPD Test

The traditional tests for selecting a distribution model, for example, Kolmogorov-Smirnov or Cramèr-von Mises, validate the calibration of the model only for the central part of the distribution of observations. Certain versions of the Cramèr-von Mises test even try to reduce the influence of extreme observations through an adapted weighting pattern giving the same importance to each of the observations. Several other tests and calibration criteria are frequently used, each one impacting on the entire sample.

However, in the context of frequency analysis, we are concerned above all with the estimation of extreme events, and so we must focus more on the calibration of the selected model for the tails of the distributions. Unfortunately, it is usually difficult to meet this objective because we very rarely, if not to say never, have access to enough data that corresponds to the extreme events that we are trying to estimate.

A calibration test was recently developed (Garrido, 2002) in connection with the estimation of extreme events. This test, the so-called GPD test, is intended to examine the goodness-of-fit of a parametric model dealing with extreme values.

The goal of the procedure is to test the calibration of a parametric model by concentrating on the tail of the distribution (Fig. 4.4 (a)) and by giving more weight to the extreme observations (Fig. 4.4 (b)).

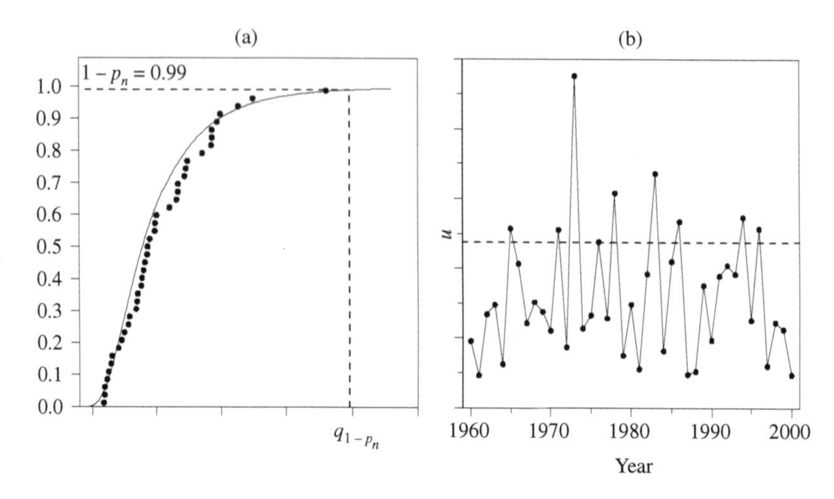

Fig. 4.4 Illustration of the goal of the procedure: (a) Empirical and calibrated distributions, (b) Exceedance model (taken from Restle *et al.*, 2004).

The theoretical basis of this approach is that the excesses above a threshold u can be approximated by the generalized Pareto distribution. The basic idea is summarized in the following steps:

1. estimate the quantile q_{1-p_n} of order 1-p_n by using the excesses above u and the approximation from the Pareto distribution;
2. construct a confidence interval CI for q_{1-p_n} based on this estimator;
3. accept the parametric model if the parametric estimator $\hat{q}_{\text{param},n}$ lies within the CI (confidence interval).

Step 2 is carried out with a parametric bootstrap. The details of performing this test can be found in Restle *et al.* (2004). This report examines four parametric models: the generalized extreme values distribution, the three-parameter lognormal distribution, the Pearson type III distribution and the Halphen type A distribution (Morlat, 1956). The goal is to examine by means of simulation the dependence between the level of the test and the model parameters in order to use the same value for threshold u independently of the values of the estimated parameters. This is in fact a limitation of the method because the threshold has to be determined through simulations.

4.2.5 Diagrams of Moments

Moments or centered moments are used to characterize the distribution of a random variable $X \sim F(X)$. The uncentered moment of order r is expressed by

$$\mu_r' = E(X^r) = \int_{-\infty}^{\infty} x^r f(x)dx \tag{4.3}$$

For example, inserting $r = 1$ in equation (4.3), one obtains $\mu_1' = E(X) = \mu$.

The centered moment of order r is obtained by subtracting its expectation from x in equation (4.3). This gives us

$$\mu_r = \int_{-\infty}^{\infty} (x - \mu_1')^r f(x)dx \tag{4.4}$$

The case $r = 2$ corresponds to the most commonly used centered moment, namely variance. In essence

$$\mu_2 = \int_{-\infty}^{\infty} \{x - E(x)\}^2 f(x)dx$$

which is nothing more than the definition of variance σ^2.

The relationship between centered and uncentered moments can be found in Kendall *et al.* (1999), among others, such as the following:

$$\begin{aligned} \mu_r &= \sum_{j=0}^{r} \binom{r}{j} \mu_{r-j}' (-\mu_1')^j \\ \mu_r' &= \sum_{j=0}^{r} \binom{r}{j} \mu_{r-j} (\mu_{r-1}')^j \end{aligned} \tag{4.5}$$

For example by using equation (4.5), the centered moment of order 3 can be expressed as

$$\begin{aligned} \mu_3 &= \binom{3}{0} \mu_3' + \binom{3}{1} \mu_2' (-\mu_1') + \binom{3}{2} \mu_1' (-\mu_1')^2 + \binom{3}{3} \mu_0' (-\mu_1')^3 \\ &= \mu_3' - 3\mu_2' \mu_1' + 2\mu_1'^3 \end{aligned} \tag{4.6}$$

In statistical hydrology, the main moments ratio employed are the coefficient of variation (C_v), the coefficient of asymmetry (C_s) and the coefficient of flatness (or kurtosis) (C_k), defined respectively by:

$$\begin{aligned} C_v &= \frac{\sqrt{\mu_2}}{\mu_1'} = \frac{\sigma}{\mu} \\ C_s &= \gamma_1 = \frac{\mu_3}{\mu_2^{3/2}} = \frac{\mu_3}{\sigma^3} \\ C_k &= \gamma_2 = \frac{\mu_4}{\mu_2^2} - 3 = \frac{\mu_4}{\sigma^4} - 3 \end{aligned} \tag{4.7}$$

Similarly to the theoretical moments above, it is possible to define sample moments. Uncentered r order sample moments are obtained as:

$$m'_r = \frac{1}{n}\sum_{i=1}^{n} x_i^r \tag{4.8}$$

Centered sample moments of order r are computed by subtracting the mean of the sample in equation (4.7), leading to

$$m_r = \frac{1}{n}\sum_{i=1}^{n} (x_i - \bar{x})^r \tag{4.9}$$

These moments are often biased and can be partly corrected in the following way:

$$\begin{aligned} \hat{\mu}_2 &= \frac{1}{n-1}\sum_{i=1}^{n}(x_i - \bar{x})^2 \\ \hat{\mu}_3 &= \frac{n}{(n-1)(n-2)}\sum_{i=1}^{n}(x_i - \bar{x})^3 \\ \hat{\mu}_4 &= \frac{n^2}{(n-1)(n-2)(n-3)}\sum_{i=1}^{n}(x_i - \bar{x})^4 \end{aligned} \tag{4.10}$$

However in small samples the bias cannot be removed.

The sample coefficients of asymmetry c_s and flatness c_k are expressed by:

$$c_s = \frac{m_3}{m_2^{3/2}} \qquad c_k = \frac{m_4}{m_2^2} - 3 \tag{4.11}$$

Several corrections of bias can be found in the literature, for example:

$$\begin{aligned} (c_s)_1 &= \frac{\sqrt{n(n-1)}}{n-2}\, c_s \ \text{(WRC, 1967)} \\ (c_s)_2 &= (c_s)_1\left(1 + \frac{8.5}{n}\right) \ \text{(Hazen, 1924)} \\ (c_s)_3 &= \left[\left(1 + \frac{6.51}{n} + \frac{20.2}{n^2}\right) + \frac{1.48}{n} + \frac{6.77}{n^2}c_s^2\right] c_s \end{aligned} \tag{4.12}$$

(Bobée and Robitaille, 1975)

It is obvious that when n is large, $(c_s)_1$, $(c_s)_2$ and $(c_s)_3$ tend towards c_s.

For a given distribution, the theoretical moments are functions of the distribution parameters. Moments of larger order can be written like the equations for lower order moments. For some three-parameter distributions, C_k can be expressed as a function of C_s. The idea of a diagram of moments is to represent C_k as a function of C_s for some of the distributions used in frequency analysis and to locate the point (c_s, c_k) obtained based on the sample on the same graphic. The formulas below express the relation between C_k and C_s for some common distributions.

Normal distribution (N): $C_k = 0,\ C_s = 0$

three-parameter lognormal distribution (LN3):

$$C_k = 3 + 0.0256C_s + 1.7205C_s^2 + 0.0417C_s^3 + 0.0460C_s^4 - 0.0048C_s^5 + 0.0002C_s^6$$

Pearson type III distribution (P3): $C_k = 3 + 1.5C_s^2$

Generalized extreme-value distribution (GEV):

$$C_k = 2.6951 + 0.1858C_s + 1.7534C_s^2 + 0.1107C_s^3 + 0.0377C_s^4 + 0.0036C_s^5 + 0.0022C_s^6 + 0.0007C_s^7 + 0.00005C_s^8$$

Figure 4.5 shows the obtained curves. It should be noted that two-parameter distributions are represented with a point, whereas three-parameter distributions correspond to curves.

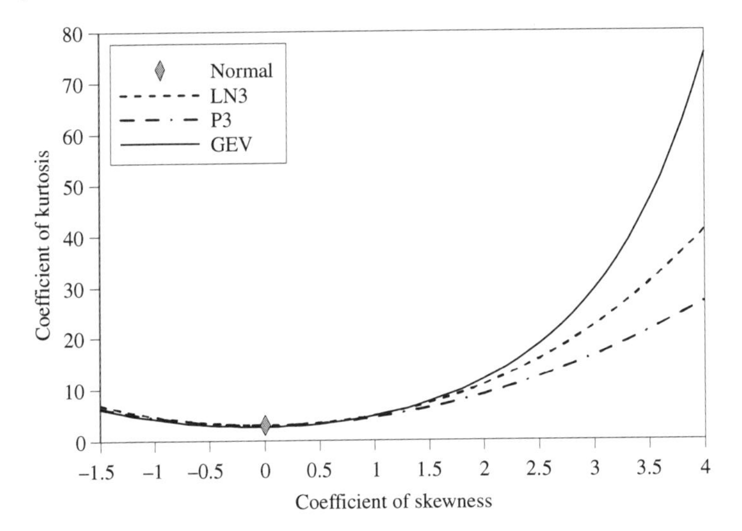

Fig. 4.5 Diagram of moments for normal, three-parameter lognormal (LN3), Pearson type III (P3) and generalized extreme-value (GEV) distributions.

As an example of the application of a diagram of moments, let us look at the case of the annual maximum daily discharges of rivers in Quebec. The criterion for retaining a station is that it has at least 20 years of data. This means that 176 stations were retained in this example. The corresponding coefficients of asymmetry and flatness are represented by a point in the diagram of moments, as shown in Figure 4.6.

This type of diagram makes it possible in theory to test the goodness-of-fit of the distribution chosen for a station, as well as the regional homogeneity. We can observe that most of the stations are located in the portion where the curves corresponding to the three-parameter distributions are indistinguishable. For about twenty of the stations, the Pearson type III distribution is preferable. Clearly, the most suitable distribution for the whole set of stations is the Pearson type III.

The two main limits to this method are the facts that typically the sample moments are biased for small samples, and that the curves representing the various distributions used in hydrology (not all of which are included in Figure 4.6) are relatively confined, which makes it quite difficult to provide a clear differentiation.

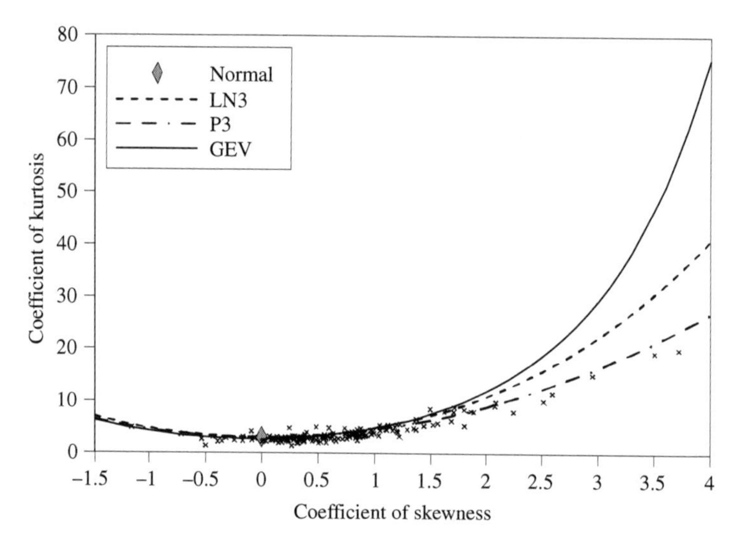

Fig. 4.6 Diagram of moments for normal, lognormal with three parameters (LN3) and Pearson type III (P3) distributions and the generalized extreme-value (GEV) distribution. The x-marks correspond to the coefficients of asymmetry and flatness for the annual maximum discharges for 176 stations in the province of Quebec.

4.2.6 Diagram of L-moments

Probability weighted moments (PWMs) can be defined by (Greenwood *et al.*, 1979):

$$M_{p,r,s} = \mathrm{E}\{x^p F^r (1-F)^s\} = \int_0^1 \{x(F)\}^p \, F^r (1-F)^s \, dF \tag{4.13}$$

The most often used probability weighted moments are obtained by setting particular values for p, r and s. For example:

$$p = 1,\ r = 0 \ \rightarrow M_{1,0,s} = \alpha_s = \int_0^1 x(F)(1-F)^s \, dF$$

$$p = 1,\ s = 0 \ \rightarrow M_{1,r,0} = \beta_r = \int_0^1 x(F) F^r \, dF \qquad (4.14)$$

$$r = s = 0 \qquad \rightarrow M_{p,0,0} = \mu'_r$$

There is a link between α_s and β_r which can be expressed as:

$$\alpha_s = \sum_{i=0}^{s} \binom{s}{i} (-1)^i \beta_i$$

$$\beta_r = \sum_{i=0}^{r} \binom{r}{i} (-1)^i \alpha_i \qquad (4.15)$$

By using equation (4.15), for example, we obtain

$$\alpha_0 = \beta_0$$

$$\alpha_1 = \beta_0 - \beta_1 \qquad \beta_1 = \alpha_0 - \alpha_1 \qquad (4.16)$$

$$\alpha_2 = \beta_0 - 2\beta_1 + \beta_2 \quad \beta_2 = \alpha_0 - 2\alpha_1 + \alpha_2$$

The sample probability moments corresponding to α_s and β_r are computed as follows. Supposing that $x_1 \leq \cdots \leq x_n$ and $n > r$, $n > s$:

$$a_s = \hat{\alpha}_s = \hat{M}_{1,0,s} = \frac{1}{n} \frac{\sum_{i=1}^{n} \binom{n-i}{s} x_i}{\binom{n-1}{s}}$$

$$b_r = \hat{\beta}_r = \hat{M}_{1,r,0} = \frac{1}{n} \frac{\sum_{i=1}^{n} \binom{i-1}{r} x_i}{\binom{n-1}{r}} \qquad (4.17)$$

These estimators have the advantage of being unbiased.

Hosking (1990) defines L-moments in the following way:

$$\lambda_{r+1} = (-1)^r \sum_{k=0}^{r} p^*_{r,k} \alpha_k = \sum_{k=0}^{r} p^*_{r,k} \beta_k \qquad (4.18)$$

where

$$p^*_{r,k} = (-1)^{r-k} \binom{r}{k} \binom{r+k}{k}$$

For example, by using equation (4.18), we obtain

$$\lambda_1 = \alpha_0$$

$$\lambda_2 = \alpha_0 - 2\alpha_1 \qquad \lambda_1 = \beta_0$$

$$\lambda_3 = \alpha_0 - 6\alpha_1 + 6\alpha_2 \qquad \lambda_2 = -\beta_0 + 2\beta_1 \qquad (4.19)$$

$$\lambda_3 = \alpha_0 - 6\alpha_1 + 6\alpha_2 \qquad \lambda_3 = \beta_0 - \beta_1 + 6\beta_2$$

$$\lambda_4 = \alpha_0 - 12\alpha_1 + 30\alpha_2 - 20\alpha_3 \qquad \lambda_4 = -\beta_0 + 12\beta_1 - 30\beta_2 + 20\beta_3$$

Similarly to the relation used for moments, Hosking (1990) defined the L-moments ratio as:

$$\tau = \frac{\lambda_2}{\lambda_1}$$

$$\tau_r = \frac{\lambda_r}{\lambda_2}, \; r \geq 3 \tag{4.20}$$

where λ_1 is a parameter of location, τ is a parameter of scale and of dispersion (LC_v), τ_3 is a parameter of symmetry (LC_s) and τ_4 is a parameter of flatness or kurtosis (LC_k).

The diagram of L-moments is based on the same principle as the one for moments. The idea is to represent the L-coefficient of flatness as a function of the L-coefficient of asymmetry.

Figure 4.7 illustrates a diagram of L-moments for normal, three-parameter lognormal (LN3), Pearson type III (P3), and generalized extreme-value (GEV) distributions. The curve corresponding to the lower limit is also shown on the graph.

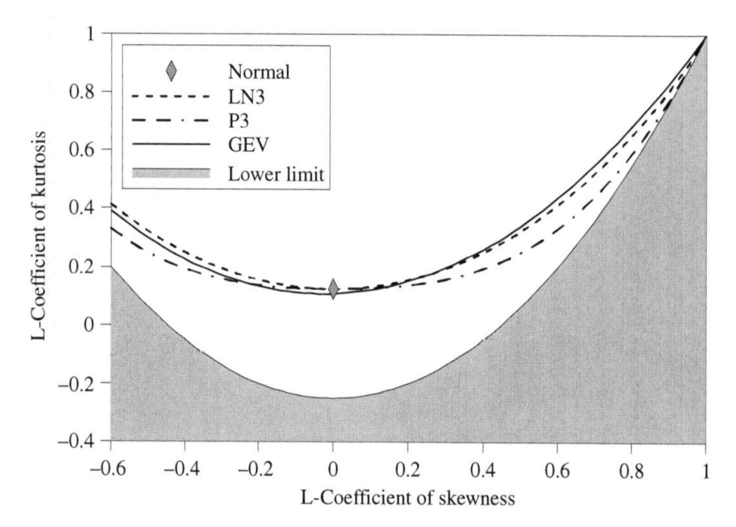

Fig. 4.7 L-moments diagram of normal, three-parameter lognormal (LN3), Pearson type III (P3) and generalized extreme-value (GEV) distributions. The lower limit (BI) is also represented by a curve.

For an application, we used the same 176 annual maximum discharge series from the province of Quebec. Figure 4.8 illustrates the obtained results.

L-moments result in a unified approach for a complete statistical inference (Hosking, 1990). Vogel and Fennessey (1993) demonstrated

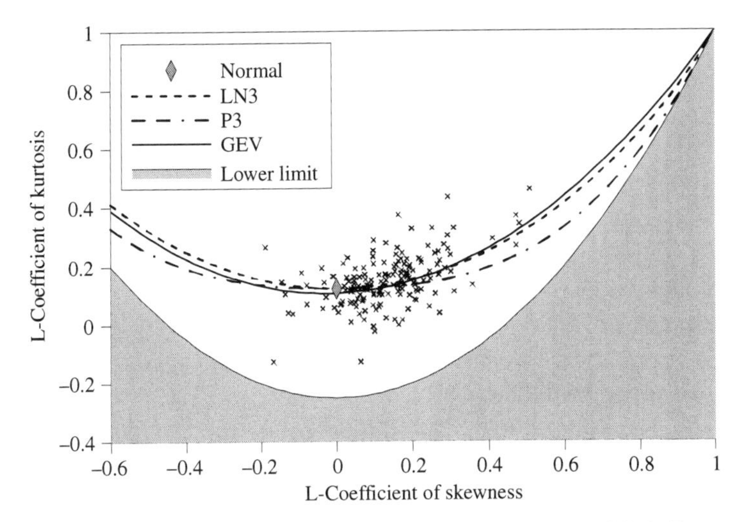

Fig. 4.8 L-moments diagram of normal, three-parameter lognormal (LN3), Pearson type III (P3) and generalized extreme-value (GEV) distributions. The x-marks correspond to the coefficient of asymmetry and flatness for the annual maximum discharge at 176 stations from the province of Quebec.

the advantages of L-moments diagrams and in particular, the fact that L-moments allow a better differentiation of the various distributions while being highly robust. They also emphasized that L-moments suffer less from the effects of sample variability. They are also less sensitive to the presence of outliers.[17]

However Bernier (1993) pointed out that robustness in the face of high values can eliminate important information contributed by extreme values. Ben-Zvi *et al.* (1997) noted that the L-moments diagram does not allow to determine the most appropriate of the possible distributions. Klemeš (2000) stressed that this diagram seems to lead in numerous cases to choosing the generalized extreme-value (GEV) distribution.

4.3 CONCLUSIONS

The various aspects of choosing a frequency model discussed throughout this chapter demonstrate that the hydrologist remains rather impoverished in the face of this problems. Consequently, it is not possible to recommend any more precise rules than the very general principles formulated below:

[17] It must be stressed that an outlier in the statistical sense does not imply that the value is false or lacks physical meaning.

1. The choice of a model cannot be based exclusively on mathematical considerations; supplementary information is required, objective if possible, but often and in practice, subjective – expert knowledge, experience with similar cases, similarities with other phenomena, etc.
2. Given the considerations outlined above (Sections 4.2.5 and 4.2.6) the "traditional" diagrams of moments should no longer be used. Replacing these with L-moments diagrams will help greatly in choosing the appropriate model.
3. Goodness-of-fit tests are not useful in choosing between several models; they can only confirm the goodness-of-fit of a model given the available sample.
4. A recent test, the GPD, is appropriate for frequency analysis. However the theoretical framework requires further development (for example regarding the power of the test).
5. Whatever the method adopted and its statistical performance, it is wise to remember that the representativeness of the series of values used is linked only to the duration (in years) of the series of observations.

CHAPTER 5

Estimation of Model Parameters

In this chapter we will examine the techniques for the *calibration* or *fitting*, or sometimes called *specification*, of a frequency model for a data series: this consists, essentially, of estimating the parameters of the model. We will look at both classical statistical techniques for parameter estimation, and methods used specifically in frequency hydrology.

5.1 THE MAXIMUM LIKELIHOOD METHOD

Given a sample $(x_1,x_2,\cdots,x_n)$ of independent identically distributed observations following a distribution F_θ where $\theta = (\theta_1,\theta_2,\cdots,\theta_p)$ is the p-dimensional vector of the parameters of the distribution, then by definition, the probability density is given by:

$$\Pr(x_i \leq X \leq x_i + dx_i) = f_\theta(x_i)dx_i$$

As the observations are independent and identically distributed following a distribution F_θ, it follows that the probability p_1 of observing a sample $(x_1,x_2,\cdots,x_n)$ can be written as:

$$p_1 = f_\theta(x_1)dx_1 f_\theta(x_2)dx_2 \cdots f_\theta(x_n)dx_n \tag{5.1}$$

The likelihood is thus expressed as:

$$V(\theta) = f_\theta(x_1)f_\theta(x_2)\cdots f_\theta(x_n) = \prod_{i=1}^{n} f_\theta(x_i) \tag{5.2}$$

By comparing equations (5.1) and (5.2) we can see that the probability p_1 is proportional to the likelihood V.

The maximum likelihood method consists of finding the vector of parameters $\hat{\theta} = (\hat{\theta}_1,\hat{\theta}_2,\cdots,\hat{\theta}_p)$ that maximizes the likelihood and therefore the probability of observing the sample $(x_1,x_2,\cdots,x_n)$.

The maximum of the likelihood function is obtained by solving the following system of p equations in p unknowns:

$$\frac{\partial V(\theta)}{\partial \theta_j} = 0, \quad j = 1,\cdots,p \tag{5.3}$$

For practical reasons, we often seek to maximize the logarithm of the likelihood, often called the log-likelihood. This leads to the transformation of the product of equation (5.2) to a sum.

As an example, we will compute the maximum likelihood estimator for a Gamma distribution with two parameters (α and λ). The Gamma probability density function is written as:

$$f(x) = \frac{\alpha^{\lambda}}{\Gamma(\lambda)}\exp(-\alpha x)x^{\lambda-1}$$

The likelihood can therefore be expressed by:

$$V = \prod_{i=1}^{n} f(x_i) = \frac{\alpha^{\lambda n}}{\{\Gamma(\lambda)\}^n}\exp\left(-\alpha\sum_{i=1}^{n} x_i\right)(x_1 x_2 \cdots x_n)^{\lambda-1} \tag{5.4}$$

It is simpler in this case to maximize the log-likelihood which is written as:

$$\ln(V) = n\lambda\ln(\alpha) - n\ln\Gamma(\lambda) - \alpha\sum_{i=1}^{n} x_i + (\lambda - 1)\sum_{i=1}^{n} \ln x_i \tag{5.5}$$

Parameter estimation consists of finding $\hat{\alpha}$ and $\hat{\lambda}$ that maximize equation (5.4) or (5.5). In the case of a system of 2 equations in 2 unknowns, this can be written as:

$$\begin{cases} \dfrac{\partial \ln V}{\partial \alpha} = \dfrac{n\lambda}{\alpha} - \displaystyle\sum_{i=1}^{n} x_i \\ \dfrac{\partial \ln V}{\partial \lambda} = n\ln\alpha - n\dfrac{\partial \ln\Gamma(\lambda)}{\partial \lambda} \end{cases} \tag{5.6}$$

By substituting $A = \frac{1}{n}\sum_{i=1}^{n} x_i$, $G = (x_1 x_2 \cdots x_n)^{1/n}$ and $\frac{\partial \ln\Gamma(\lambda)}{\partial \lambda} = \psi(\lambda)$, the system of equations (5.6) can be written in the following form:

$$\begin{cases} n\dfrac{\lambda}{\alpha} - nA = 0 \\ n\ln\alpha - n\psi(\lambda) + n\ln G = 0 \end{cases} \Rightarrow \begin{cases} \dfrac{\lambda}{\alpha} = A \\ \psi(\lambda) - \ln\alpha = \ln G \end{cases} \tag{5.7}$$

For the solution, in practice we would replace α with its value in the second equation:

$$\begin{cases} \alpha = \dfrac{\lambda}{A} \\ \psi(\lambda) - \ln \lambda = \ln G - \ln A \end{cases} \tag{5.8}$$

If we denote $U = \ln G - \ln A$ then an approximated value $\hat{\lambda}_a$ *of* λ is given by:

$$\hat{\lambda}_a = \frac{1 + \sqrt{1 + 4U/3}}{4U} \tag{5.9}$$

A more precise approximation of λ was proposed by Bobée and Desgroseillers (1985) as follows:

$$\hat{\lambda} = \hat{\lambda}_a - \Delta\hat{\lambda}_a = \hat{\lambda}_a - 0.04475(0.26)^{\hat{\lambda}_a} \tag{5.10}$$

5.2 THE METHOD OF MOMENTS

The method of moments simply consists of equalling theoretical uncentered moments μ'_r and centered moments μ_r with their empirical estimations (m'_r and m_r respectively) based on the available sample. We need to consider at least as many independent moments as there are parameters to be estimated. In general moments of the lowest possible order should be used. For example for 2-parameter distributions, the mean (μ'_1) and the variance (μ_2) should be used. It should be noted that for 3-parameter distributions, the estimation should not be based on the second order uncentered moment μ'_2 and its centered counterpart because $\mu_2 = \mu'_2 - \mu'^2_1$ and in such case the chosen moments are not independent.

As an example, we are going to estimate the parameters of the Gumbel distribution based on the method of moments. The Gumbel distribution can be written as:

$$F(x) = \exp\left\{- \exp\left(- \frac{x - \alpha}{\beta} \right)\right\} \tag{5.11}$$

where α and β are the parameters to be estimated. The mean and the variance of a random variable following a Gumbel distribution are expressed as a function of parameters α and β as follows:

$$\begin{cases} \mu'_1 = \mu = \alpha + \beta C \\ \mu_2 = \sigma^2 = \beta^2 \dfrac{\pi^2}{6} \end{cases} \tag{5.12}$$

where $C = 0.5772$ denotes Euler-Mascheroni constant.

The method of moments results in the following system of 2 equations in 2 unknowns:

$$\begin{cases} \mu_1' = m_1 = \bar{x} \\ \mu_2 = m_2 \end{cases} \Rightarrow \begin{cases} \hat{\beta} = \frac{\sqrt{6}}{\pi} m_2 \\ \hat{\alpha} = \bar{x} - \hat{\beta}C \end{cases} \tag{5.13}$$

There are several techniques related to the method of moments, for example, the indirect method of moments based on the sample of the logarithm of the observations, the direct method of moments developed by Bobée (1975) for the log-Pearson type III distribution, the mixed method of moments (Rao, 1983) which is an extension of the direct method of moments, and the Sundry Average Method (SAM) (Bobée and Ashkar, 1988).

Example

Let us look again at the annual maximum mean daily discharge series from the Massa River at Blatten (see Table 1.1).

The estimator of μ for the population is given by the arithmetic mean of the 80 values (also called sample mean or empirical mean) building the sample as follows:

$$\hat{\mu} = \bar{x} = 73.46$$

The standard deviation σ of the population is computed as the sample standard deviation of n=80 values:

$$\hat{\sigma} = \sqrt{m_2}\sqrt{\frac{n}{n-1}} = \sqrt{\frac{1}{n-1}\sum_{i=1}^{n}(x_i - \bar{x})^2} = 11.75$$

It should be noted here that, as mentioned in Section 4.2.5,

$$m_2 = \frac{1}{n}\sum_{i=1}^{n}(x_i - \bar{x})^2$$

is a biased estimator of the variance. It is therefore better to use

$$s^2 = \frac{1}{n-1}\sum_{i=1}^{n}(x_i - x)^2$$

which is unbiased.

Using the equations (5.13) we therefore obtain:

$$\begin{cases} \hat{\beta} = \frac{\sqrt{6}}{\pi} 11.75 = 9.16 \\ \hat{\alpha} = 73.46 - 9.16 \times 0.5772 = 68.17 \end{cases}$$

5.3 THE METHOD OF PROBABILITY WEIGHTED MOMENTS

The method of probability weighted moments was developed by Landwehr and Matalas (1979). This method is based on the same principle as the (conventional) method of moments and consists of equalling the theoretical weighted moments β_r (or α_s) and their estimations b_r (or a_s) based on the sample. For a 3-parameter distribution, this leads to solving the system $\beta_i = b_i$, $i = 0, 1, 2$.

This method is especially interesting if, based on $F(x)$, an analytical expression $x(F)$ can be deduced because in this case we can obtain β_r directly as follows:

$$\beta_r = \int_0^1 x(F)\ \{F(x)^r\}\ dF \tag{5.14}$$

In the case of the Gumbel distribution, F can be inverted to find the expression for x, so that:

$$-\ln(-\ln F) = \frac{x - \alpha}{\beta} \Rightarrow x(F) = \alpha - \beta \ln(-\ln F) \tag{5.15}$$

By replacing x with its value in equation (5.14) we obtain:

$$\beta_r = \int_0^1 \{\alpha - \beta \ln(-\ln F)\} F^r dF \tag{5.16}$$

Greenwood *et al.* (1979) showed that the solution of this integral is:

$$\beta_r = \frac{\alpha}{1 + r} + \frac{\{\ln(1 + r) + C\}}{1 + r} \tag{5.17}$$

with $C = 0.5772$ (Euler-Mascheroni constant)

Now we have

$$\begin{cases} \beta_0 = \alpha + \beta C \\ \beta_1 = \frac{1}{2}\{\alpha + \beta(\ln 2 + C)\} \end{cases} \tag{5.18}$$

By replacing β_0 and β_1 with their empirical estimations b_0 and b_1 based on the sample, we obtain the following system of 2 equations in 2 unknowns:

$$\begin{cases} b_0 = \alpha + \beta C \\ b_1 = \frac{1}{2}\{\alpha + \beta\ (\ln 2 + C)\} \end{cases} \Rightarrow \begin{cases} \hat{\beta} = \dfrac{2b_1 - b_0}{\ln 2} \\ \hat{\alpha} = b_0 - \hat{\beta} C \end{cases} \tag{5.19}$$

with

$$b_0 = \bar{x} \text{ et } b_1 = \sum_{i=1}^{n} \left(\frac{i - 1}{n - 1}\right) x_{[i]} \tag{5.20}$$

5.4 THE METHOD OF *L*-MOMENTS

As we have already seen, *L*-moments are linear combinations of probability weighted moments (see Section 4.2.6). As a result, the method of *L*-moments and the method of probability weighted moments can be considered as equivalent. For the Gumbel distribution the two estimators $\hat{\beta}$ and $\hat{\alpha}$ can be obtained very easily based on the values of the first two *L*-moments of a Gumbel population and their corresponding estimations computed on the sample. This leads to:

$$\begin{cases} \hat{\beta} = \dfrac{\hat{\lambda}_2}{\ln 2} \\ \hat{\alpha} = \hat{\lambda}_1 - 0.5772\,\hat{\beta} \end{cases} \tag{5.21}$$

It should be mentioned here that research in the field of parameter estimation using the method of *L*-moments is still ongoing (El Adlouni *et al.*, 2007).

5.5 THE GRAPHICAL METHOD

A graphical method can be used whenever the expression for the quantile corresponds to the equation for a straight line. After linearization, many 2-parameter distributions satisfy this criterion, (normal distribution, exponential distribution, and the Gumbel distribution, for example). The graphical method consists mainly of entering points representing the observed values on a graph *linearizing* the relation between the quantile x and the *cumulative frequency* F: this type of graphic is called a *probability paper.*

5.5.1 Application of the Gumbel Distribution

The principle

The graphical method is based on the fact that in the case of a Gumbel distribution, the expression for a quantile corresponds to the equation for a *straight line.* In fact, by inserting the reduced variable $u = \dfrac{x - \alpha}{\beta}$ in the expression for the Gumbel distribution (equation 5.11), we obtain $x = \alpha + \beta u$.

Thus when the points representing the series to be calibrated are plotted on the axes x-u, it becomes possible to draw a straight line that at best fits these points and to deduce the two parameters $\hat{\alpha}$ and $\hat{\beta}$ of the distribution (see Fig. 5.1).

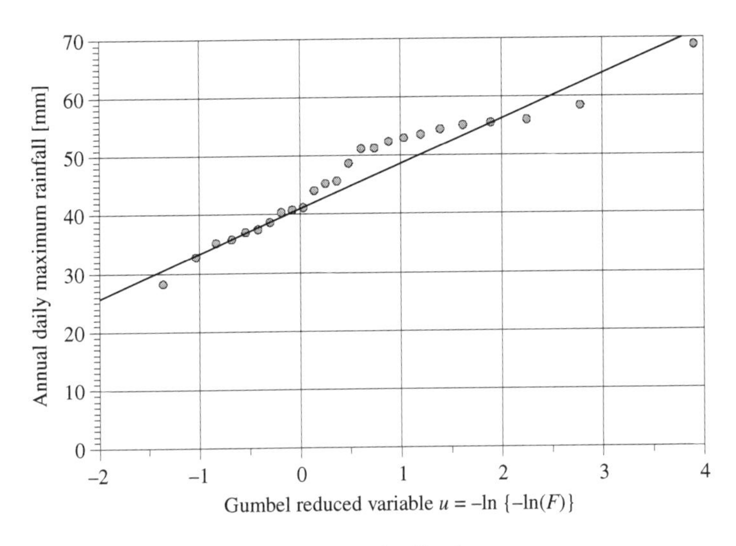

Fig. 5.1 Principle of the graphical method of calibration.

Assuming that the points x_i are known (they belong to the available data), it can be useful to define the coordinates u_i corresponding to each of these points in order to plot them on the graph.

These coordinates are determined based on the inverse relation of the distribution function which gives u as a function of the cumulative frequency $F(x)$. This leads mainly to estimating the probability of nonexceedance $F(x_i)$ that can be attributed to each value x_i.

A Reminder of Useful Equations for the Graphical Method

For the Gumbel distribution, these relations can be deduced from equation (5.11) as follows:

$$F(x) = \exp\{-\exp(-u)\};\ u = -\ln\{-\ln(F(x)\};\ u = \frac{x-\alpha}{\beta};\ x = \alpha + \beta u$$

Empirical Frequency

As we have already seen in Chapter 1, there are a number of formulas for the estimation of the cumulative frequency $F(x)$, including *plotting position* formulas and formulas for estimating the *empirical frequency*. All of them are based on the sorting of the series by ascending (or descending) order, making it possible to associate a rank of order r for each value. In practice, these formulas can be summarized in the following general equation that ensures symmetry around the median:

$$\hat{F}(x_{[r]}) = \frac{r-c}{n+1-2c} \tag{5.22}$$

where n denotes the sample size, $x_{[r]}$ denotes the value corresponding to the rank r and c is a coefficient taking values between 0 and 0.5. Table 5.1 shows some examples of empirical frequency formulas as a function of the constant c and the empirical return period corresponding to the largest value (maximum) in the sample ($x_{[n]}$).

Table 5.1 Empirical frequency formulas (adapted from Stedinger *et al.*, 1993).

Name	Formula $\hat{F}(x_{[r]})$	c	$T(x_{[n]})$	Remark
Weibull	$\frac{r}{n+1}$	0	$n + 1$	Generally used in *the* USA
Median	$\frac{r - 0.3175}{n + 0.365}$	0.3175	$1.47n + 0.5$	
Hosking	$\frac{r - 0.35}{n}$	-	$1.54n$	Asymmetric, used in the method of probability weighted moments
Blom	$\frac{r - 0.375}{n + 0.25}$	0.375	$1.6n + 0.4$	
Cunnane	$\frac{r - 0.40}{n + 0.2}$	0.40	$1.67n + 0.3$	
Gringorten	$\frac{r - 0.44}{n + 0.12}$	0.44	$1.79n + 0.2$	
Hazen	$\frac{r - 0.5}{n}$	0.5	$2n$	Mainly used in France

Theoretically there is a different empirical frequency formula for each distribution, but asymptotically all the formulas are equivalent. Several authors have tried to find which formula is best suited to a particular distribution.

Gumbel (1958, page 32) considered that in order to obtain a formula for computing empirical frequency, the following five main properties are required:

1. *"The plotting position should be such that all observations can be plotted."*
2. *"The plotting position should lie between the observed frequencies r-1/n and r/n and should be universally applicable, i.e it should be distribution-free. This excludes the probabilities at the mean, median and modal rth value which differ for different distribution."*
3. *"The return period of a value equal to or larger than the largest observation should approach n, the number of observations. This condition needs to be fulfilled by the choice of the mean or median rth value."*
4. *"The observations should be equally spaced on the frequency scale, i.e, the difference between the plotting position of the (r + 1)th and the rth observation should be a function of n only, and independent of r."*

5. *"The plotting position should have an intuitive meaning, and ought to be analytically simple. The probabilities at the mean, modal, or median rth value have an intuitive meaning."*

He concluded by advocating the Weibull formula, which is widely used in North America.

Following a theoretical study, Brunet-Moret (1973) proposed using:

- $c = 0.3$ when the parent population is *perfectly known*, which in practice is never the case;
- $c = 0.5$ (Hazen) when the parameters of the parent population are *unknown*.

Cunnane (1978), meanwhile, carried out a review of the existing formulas and concluded that the choice is related to the parent population under consideration. He also concluded that the condition that Gumbel had imposed on the order of magnitude of the return period corresponding to the largest value had no statistical basis: the probability of exceedance of the largest value in n years is 0.632[18] for a return period of $T = n$. Consequently he recommended:

- Blom ($c = 0.375$) for a normal distribution;
- Gringorten ($c = 0.44$) for a Gumbel or exponential distribution;
- Weibull ($c = 0$) for a uniform distribution.

Finally, he proposed that a coefficient of $c = 0.4$ is the "best compromise" as a *nonparametric formula*!

Based on a study of simulations, our own recommendation is to follow Brunet-Moret and adopt $c = 0.5$, which corresponds to Hazen's formula.

5.5.2 An Example of Implementation

Table 5.2 showing the annual maximum daily rainfalls at the station of Bex (Switzerland) from 1980 to 2004 will be used here as an example of the steps involved in this method:

- The time series under consideration is sorted in ascending order (column 3). This column will be the "x" values to be used.
- Column 5 simply contains the rank r corresponding to each value.
- Then the estimations $\hat{F}(x_{[r]})$ (column 6) are computed (here using Hazen's formula).
- The u coordinates, or the Gumbel standardized variable, are shown in column 7.

[18] See Section 2.2.1

Table 5.2 Example of a computation using the graphical method. Annual maximum daily rainfall at the Bex station (Switzerland) (N° 7860) from 1980 to 2004.

Chronological		Sorted				
(1) Year	(2) Daily max	(3) Daily max	(4) Year	(5) Rank r	(6) $\hat{F}$	(7) u
1980	55.2	28.2	1984	1	0.02	−1.36405
1981	48.7	32.8	2004	2	0.06	−1.03440
1982	51.3	35.2	1988	3	0.10	−0.83403
1983	41.2	35.8	1999	4	0.14	−0.67606
1984	28.2	37.0	2003	5	0.18	−0.53930
1985	52.4	37.5	1994	6	0.22	−0.41484
1986	56.1	38.7	1993	7	0.26	−0.29793
1987	58.5	40.4	1996	8	0.30	−0.18563
1988	35.2	40.8	1998	9	0.34	−0.07586
1989	55.6	41.2	1983	10	0.38	0.03295
1990	53.6	44.1	2001	11	0.42	0.14214
1991	68.8	45.3	2000	12	0.46	0.25292
1992	51.2	45.7	1995	13	0.50	0.36651
1993	38.7	48.7	1981	14	0.54	0.48421
1994	37.5	51.2	1992	15	0.58	0.60747
1995	45.7	51.3	1982	16	0.62	0.73807
1996	40.4	52.4	1985	17	0.66	0.87824
1997	53.0	53.0	1997	18	0.70	1.03093
1998	40.8	53.6	1990	19	0.74	1.20030
1999	35.8	54.5	2002	20	0.78	1.39247
2000	45.3	55.2	1980	21	0.82	1.61721
2001	44.1	55.6	1989	22	0.86	1.89165
2002	54.5	56.1	1986	23	0.90	2.25037
2003	37.0	58.5	1987	24	0.94	2.78263
2004	32.8	68.8	1991	25	0.98	3.90194

Finally, Figure 5.1 is built using the pairs x and u (column 3 and 7).

However, it should be mentioned that there are probability papers that are already scaled in cumulative frequency $F(x)$. It is superfluous in this case to transform the estimations $\hat{F}(x_{[r]})$ into the values u.

A straight line is drawn on the graph in such a way that it fits as closely as possible to the data points. The parameter α can then be read directly at the y-intercept (on the reduced Gumbel variable u axis) while the parameter β is obtained as the slope of the straight line.

In our example (see Fig. 5.1), this leads to the value $x = 41$ for $u = 0$, while for $u = 3$ it gives us $x = 64$. The location parameter α is consequently 41, and the scale parameter β is given by:

$$\hat{\beta} = \frac{\Delta x}{\Delta u} = \frac{64 - 41}{3 - 0} = 7.67 \tag{5.23}$$

Finally, we should add that although this method may seem a little archaic, it has the great advantage of supplying a *graphical representation* of the data and the goodness-of-fit. This constitutes an essential aspect of the judgment made regarding the fit between the distribution chosen and the data under study, no matter which calibration method is used.

This is the reason why we advocate a systematic graphical representation of calibration, even (and especially!) if this latter is produced with software.

Finally, we should mention that the graphical method has a numerical counterpart: the *method of least rectangles* introduced by Gumbel (1958) and recommended by the Guide to Hydrological Practices (WMO, 1981) among others. This method can be questionable when the sample contains a very high value, introducing nonlinearity.

5.6 A COMPARISON OF PARAMETER ESTIMATION METHODS

Once again we are completely unable to propose "the" perfectly appropriate estimation method. Again we mention that the choice of one method over another is often a matter of experience or of custom (see Section 4.2.3).

It is worth repeating something mentioned in the previous chapter (Section 4.1.3), to the effect that the validity of an analysis depends much more on the choice of model that on the applied calibration method.

We suggest that in the aim of comparing several estimation methods, it is wise to perform a simulation study on a "bench mark" using data, randomly of course, that are as close as possible to the data under study.

Nonetheless, we have listed below some brief considerations with regards to each of the methods examined in this chapter.

5.6.1 Maximum Likelihood Method

The maximum likelihood method is theoretically the most efficient approach in the sense that it produces the smallest sampling variance of the estimated parameters and consequently, the smallest variance of quantiles. However in some particular cases such as the log-Pearson type III distribution, the optimality is only asymptotic. This approach often results in biased estimators, although these can sometimes be corrected. In the case of small samples, it is sometimes impossible to obtain correct estimators, especially if there are a large number of parameters to be estimated.

5.6.2 Method of Moments

The method of moments is a simple and "natural" approach. Usually, the resulting estimators are not as good as those obtained with the maximum likelihood method. The difficulty of estimation in the case of small samples is induced by the fact that the higher-order moments are biased.

5.6.3 Method of Probability Weighted Moments

The estimators produced by the method of probability weighted moments are comparable to the ones obtained using the maximum likelihood method. For some distributions, this approach can simplify the estimation procedure. For small samples, this method can lead to better estimators than the maximum likelihood approach. Sometimes explicit expressions for the estimators are possible (the Weibull distribution, for example) which the other methods cannot achieve.

5.6.4 Graphical Methods

Graphical methods should be considered as *necessary*, first of all as an illustrative training tool, but especially as an effective way to visualize the quality and fit of the calibration (see Chapter 6).

CHAPTER 6

Validation of the Model

To begin this chapter, let us repeat one more time that a test procedure (in this case a goodness-of-fit test) only makes it possible to reject or not reject the null hypothesis $H_0 : F(x) = F_0(x)$. A goodness-of-fit test cannot be applied to *choose* the best frequency models among $F_1(x)$, $F_2(x)$, etc.

In addition, the validation of the calibration, as discussed in Section 4.1, does not constitute a validation of the model itself; it serves only as an *internal validation* of the model.

6.1 VISUAL EXAMINATION

As we have already seen in Chapter 5, a visual inspection of the graph representing the goodness-of-fit, even if it seems rudimentary, is one of the best ways to assess the quality of the fit. Figure 6.1 shows an example of such a graph.

This is clearly not a true test, because it does not provide any gradation in the probability of error. On the other hand, when a skilled operator looks at a graphic like this, it immediately summons up his or her critical judgment and synthesizing skills. Nonetheless, we must still be careful not to judge the goodness-of-fit of a model involving extreme values, given the uncertainty surrounding the empirical frequency of the largest values.

6.2 THE CHI-SQUARE TEST

The chi-square test is both a goodness-of-fit and a conformity test. In the first case, it tests whether the empirical (or experimental) cumulative frequency $\hat{F}(x)$ is not significantly different from the theoretical distribution function $F_0(x)$ resulting from the calibration. In the second case (conformity test), it compares the empirical cumulative frequency $\hat{F}(x)$ to a theoretical model $F_0(x)$ defined *a priori*.

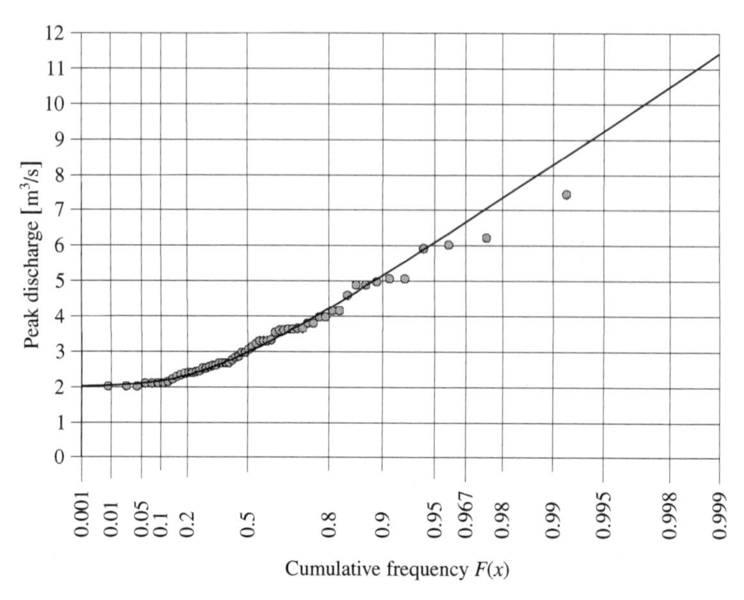

Fig. 6.1 Curve fitting of a peak over threshold (POT) series of peak discharges [m^3/s] of the Nozon River at Orny (Switzerland) (1923-1931) with an exponential distribution.

Although here we are applying the chi-square test with continuous variables, it can also be used with discrete variables.

6.2.1 Principle and Characteristics

The principle of the test involves comparing, for a given class, the observed count n_j and the theoretical count n_{0j}, which is to say, the count of class j for the theoretical distribution function $F_0(x)$. The test itself consists of verifying that the sum of the squares of the differences $\sum_j (n_j - n_{0j})^2$ stays within limits compatible with the sampling fluctuations that we would expect. In this case the null hypothesis cannot be rejected. The χ^2 test is *nonparametric.*

The null and alternative hypotheses can be expressed as follows:

$$H_0 : \hat{F}(x) = F_0(x), \ \forall x,$$

$$H_1 : \hat{F}(x) \neq F_0(x), \text{ for at least one value } x.$$

The principle of the test is illustrated in Figure 6.2 which shows the theoretical probability density f_0(x) superimposed on a normalized histogram built from the sample.

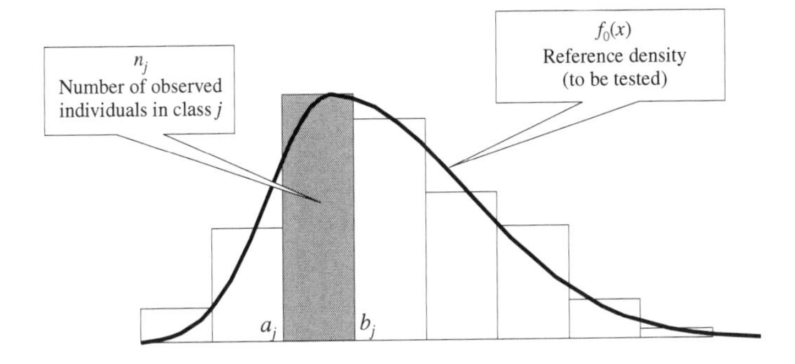

Fig. 6.2 Principle of the chi-square test.

If the sample size n is "large enough" – unfortunately the literature does not contain much more precise information! – the quantity:

$$\chi^2_{obs} = \sum_{j=1}^{k} \frac{(n_j - n_{0j})^2}{n_{0j}} \tag{6.1}$$

is distributed according to a chi-square distribution with v degrees of freedom.

The number of degrees of freedom v equals $k - 1 - p$, where k is the number of classes and p is the number of parameters of the distribution $F_0(x)$ being tested. The parameters have been estimated based on the sample. Thus, the number of degrees of freedom depends on the type of test being carried out:

- for a conformity test $v = k - 1$
- for a goodness-of-fit test $v = k - p - 1$.

Gibbons and Chakraborti, 2003 demonstrate that the theoretical count n_{0j} of each class must be at least equal to 5. Also, according to Cochran (Gibbons and Chakraborti, 2003), the power of the test is maximal when the theoretical counts n_{0j} of the classes are equal. The following procedure, which can be automated, is based on these facts.

6.2.2 Applying the Test

The procedure for applying this test involves the following steps:

1. Determine the number of classes: $k = \lfloor n/5 \rfloor$.
2. Define the limits of equiprobable classes, such that $\Delta F_j = 1/k$.
3. Compute n_{0j} (the constant) as follows: $n_{0j} = n\Delta F_j = n/k$.
4. Determine the number of the actual counts n_j of the k classes.
5. Compute the quantity $\chi^2_{\text{obs}} = \left(\frac{k}{n} \sum_{j=1}^{k} n_j^2 \right) - n$ (note that this formula is equivalent to formula (6.1) with $n_{0j} = n/k$).

6. Compute v, the number of degrees of freedom.
7. Compare the quantity χ^2_{obs} with the $(1 - \alpha)$ quantile of a chi-square distribution with v degrees of freedom or compute the p-value.
8. Conclude with the rejection or non-rejection of the null hypothesis.

6.2.3 Example

Let us consider the series of annual rainfall modules P_a measured at Lausanne from 1901 to 1979 (see Table 6.1).

The chosen distribution for this series is normal with a mean of $\hat{\mu}$ = 1065 [mm] and standard deviation of $\hat{\sigma}$ = 193 [mm].

Application of the chi-square test is carried out as follows:

1. Define the number of classes: $k = \left\lfloor \frac{n}{5} \right\rfloor = \left\lfloor \frac{79}{5} \right\rfloor = 15$.
2. Compute $\Delta F_j = 1/15 = 0.0667$.
3. Build table of cumulative frequencies for k classes (see column 2 of Table 6.2).
4. Determine the limits of the classes corresponding to the normal distribution $P_{aj} = \mu + \sigma \cdot z_j$, where z_j is the normal variable with zero mean and unit standard deviation corresponding to the cumulative frequency of class j (see column 4 in Table 6.2). Note that if the chosen distribution is not normal, it is obviously the chosen distribution that must be used instead.
5. Enter the limits of these classes in Table 6.1, which makes it possible to easily determine the actual count n_j of each of the classes (see column 5 of Table 6.2) and their squared values (see column 6).
6. Compute the sum of n_j^2: $\sum_{j=1}^{n} n_j^2 = 453$.

Table 6.1 Ordered series (increasing order) of annual rainfall modules [mm] at Lausanne 1901-1979. The **bold** lines indicate the limits of the classes built for the test.

Rank	Year	P_a	Rank	Year	P_a	Rank	Year	P_a
1	1921	537	28	1969	977	55	1958	1173
2	1906	693	29	1976	982	56	1951	1179
3	1953	737	30	1941	991	57	1914	1187
4	1949	744	31	1924	999	58	1950	1194
5	1945	775	32	1926	1004	59	1940	1195

Table 6.1 Contd. ...

Table 6.1 Contd. ...

6	1972	789
7	1904	790
8	1964	810
9	1971	831
10	1957	834
11	1942	836
12	1962	842
13	1943	847
14	1920	860
15	1947	880
16	1933	910
17	1973	912
18	1929	921
19	1932	927
20	1908	929
21	1938	936
22	1911	940
23	1903	951
24	1907	952
25	1934	957
26	1925	959
27	1946	960
33	1913	1015
34	1975	1030
35	1912	1041
36	1959	1044
37	1901	1047
38	1961	1064
39	1944	1069
40	1919	1072
41	1956	1085
42	1902	1088
43	1979	1094
44	1905	1100
45	1967	1113
46	1918	1116
47	1966	1123
48	1948	1129
49	1963	1129
50	1928	1136
51	1974	1152
52	1954	1156
53	1937	1161
54	1909	1163
60	1978	1195
61	1931	1207
62	1936	1209
63	1917	1225
64	1955	1233
65	1970	1240
66	1927	1250
67	1923	1274
68	1935	1276
69	1915	1285
70	1952	1286
71	1960	1294
72	1916	1305
73	1939	1309
74	1977	1326
75	1965	1373
76	1968	1374
77	1910	1384
78	1922	1464
79	1930	1572

7. Compute the corresponding score $\chi^2_{\text{obs}} = \frac{15}{79}453 - 79 = 7.01$.
8. Determine the number of degrees of freedom $\nu = 15 - 2 - 1 = 12$ (a normal distribution has two parameters (μ and σ) that have been estimated based on the sample).
9. Compare the score with the 95% quantile of a chi-square distribution $\chi^2_{12}(95\%) = 21.03$. As $7.01 < 21.03$, the null hypothesis H_0 cannot be rejected.

6.2.4 Remarks

The chi-square test has notoriously a weak power (meaning that it does not allow to clearly refute the null hypothesis H_0 when the alternative hypothesis H_1 is in reality true). In the case of the example given in Section 6.2.3, it would have been wiser to apply a goodness-of-fit test adapted to a normal distribution, such as Lilliefors or Shapiro-Wilks.

In addition, the grouping into classes results in a loss of information which is damaging, especially for the extreme classes.

Table 6.2 Example of implementation of the chi-square test.

J	F_j	z_j	Pa_j	n_j	n_j^2
1	0.0667	$-\infty$	$-\infty$	5	25
2	0.1333	-1.50	775.5	8	64
3	0.2000	-1.12	848.8	2	4
4	0.2667	-0.85	900.9	7	49
5	0.3333	-0.63	943.4	6	36
6	0.4000	-0.43	982.0	4	16
7	0.4667	-0.26	1014.8	5	25
8	0.5333	-0.08	1049.6	3	9
9	0.6000	+.008	1080.4	5	25
10	0.6667	+.026	1115.2	7	49
11	0.7333	+0.43	1158.0	4	16
12	0.8000	+.063	1186.6	7	49
13	0.8667	+0.85	1229.0	5	25
14	0.9333	+1.12	1281.2	6	36
15	1.0000	+1.50	1354.5	5	25
		$+\infty$	$+\infty$		

The main advantage of this test is based on the fact that it allows, through the number of degrees of freedom $\nu = k - 1 - p$, to account for the fact that $F_0(x)$ was estimated based on the available sample.

6.3 THE KOLMOGOROV-SMIRNOV TEST

6.3.1 Principle

The Kolmogorov-Smirnov test consists of measuring, for a *continuous* variable, the largest difference between the theoretical frequency $F_0(x)$ and the empirical frequency $\hat{F}(x)$. Figure 6.3 illustrates the principle of the test. Basically, this is a test of conformity rather than a goodness-of-fit test: it consists of verifying whether an experimental distribution can be considered as identical to a reference one, known *a priori*. The null and alternative hypotheses can be written as:

$$H_0 : \hat{F}(x) = F_0(x), \forall x;$$

- bilateral test: $H_1 : \hat{F}(x) \neq F_0(x)$, for at least one x;
- unilateral test: $H_1 : \hat{F}(x) > F_0(x)$ or $\hat{F}(x) < F_0(x)$, for at least one value x.

The bilateral version of this test is more often applied.

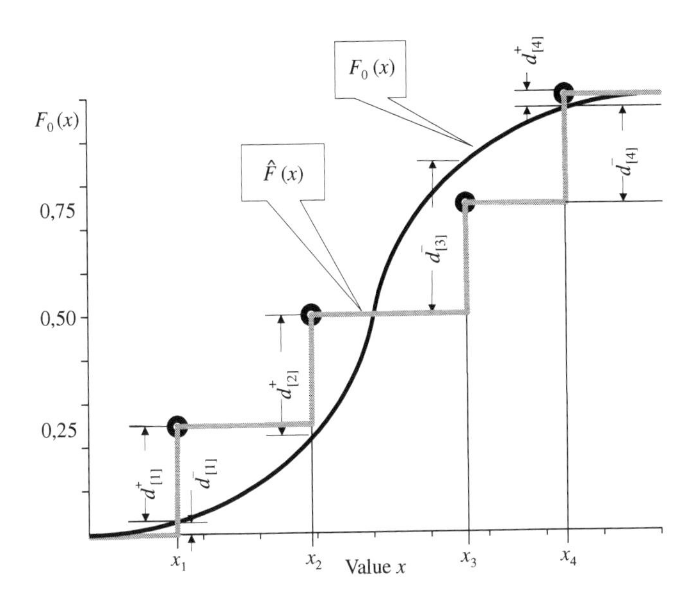

Fig. 6.3 Illustration of the principle of the Kolmogorov-Smirnov test for $n = 4$.

6.3.2 Implementation

Let us suppose that the sample of interest, with n values of x_i, is sorted in ascending order. Each observation of rank order r, $x_{[r]}$ has a corresponding empirical frequency $\hat{F}(x_{[r]})$ and a theoretical frequency $F_0(x_{[r]})$.

According to Kolmogorov-Smirnov's theory, the experimental, or observed frequency, is computed using the equation traditionally used by statisticians:

$$\hat{F}(x_{[r]}) = \frac{r}{n}$$

This means that the statistic d^+ can be defined as follows:

$$d^+ = \max\left\{\frac{1}{n} - F_0(x_{[1]}), \frac{2}{n} - F_0(x_{[2]}), \cdots, \frac{n}{n} - F_0(x_{[n]})\right\}$$

Taking into account the stair shape of the empirical frequencies diagram, the statistic d^- is defined by:

$$d^- = \max\left\{F_0(x_{[1]}) - \frac{1-1}{n}, F_0(x_{[2]}) - \frac{2-1}{n}, \cdots, F_0(x_{[n]}) - \frac{n-1}{n}\right\}$$

For a bilateral test, the statistic d is defined as the maximum of d^+ and d^-, and hence in summary:

$$\begin{cases} d^+ = \max\left\{\frac{r}{n} - F_0(x_{[r]})\right\}, \ r = 1, 2, \cdots, n \\ d^- = \max\left\{F_0(x_{[r]}) - \frac{r-1}{n}\right\}, \ r = 1, 2, \cdots, n \\ d = \max\{d^+, d^-\} \end{cases} \tag{6.2}$$

The statistic d is available in tabulated form in several works, including one by Sachs (1984, page 331) and by Siegel (1988).

For $n > 10$ Stephens (1974) gives the following formula where the coefficient $k(\alpha)$ can be found in Table 6.3:

$$d = \frac{k(\alpha)}{\sqrt{n} + 0.12 + \frac{0.11}{\sqrt{n}}} \tag{6.3}$$

Table 6.3 Coefficients for computing the Kolmogorov-Smirnov test statistic, based on Stephens (1974).

α	0.15	0.10	0.05	0.025	0.01
$k(\alpha)$	1.138	1.224	1.358	1.480	1.628

For $n > 35$, Sachs (1984) proposed the simpler, asymptotic form summarized in Table 6.4:

Table 6.4 Kolmogorov-Smirnov test statistic for $n > 35$, based on Sachs (1984).

α	0.20	0.15	0.10	0.05	0.01	0.001
$d(\alpha)$	$\frac{1.075}{\sqrt{n}}$	$\frac{1.138}{\sqrt{n}}$	$\frac{1.224}{\sqrt{n}}$	$\frac{1.358}{\sqrt{n}}$	$\frac{1.628}{\sqrt{n}}$	$\frac{1.949}{\sqrt{n}}$

6.3.3 Applied as a Goodness-of-Fit Test

The Kolmogorov-Smirnov test, which is basically a test of conformity, can also be applied to assess the goodness-of-fit. In this case the test statistic d is modified in order to take into account the reduction in the number of degrees of freedom.

Ahmad *et al.* (1988) pointed out that using the conformity test statistic for the purpose of assessing goodness-of-fit can result in non-rejection of the chosen distribution when in fact it should be rejected. They also mention that, in general, the "corrected" form of the statistic (corrected to take into account the use of the observed sample to describe the chosen distribution) can only be obtained by simulation.

In the case of a *normal distribution*, which has two parameters estimated from the observations, Stephens (1974) proposes the following formula:

$$d = \frac{k(\alpha)}{\sqrt{n} - 0.01 + \frac{0.85}{\sqrt{n}}} \tag{6.4}$$

The coefficient $k(\alpha)$ is given in Table 6.5:

Table 6.5 Coefficients for computing the Kolmogorov-Smirnov test statistic in a case of a normal distribution, based on Stephens (1974).

α	0.15	0.10	0.05	0.025	0.01
$k(\alpha)$	0.775	0.819	0.895	0.955	1.035

For a one-parameter *exponential distribution* (with a lower limit known *a priori*) we will have:

$$d = \frac{k(\alpha)}{\sqrt{n} + 0.26 + \frac{0.5}{\sqrt{n}}} + \frac{0.2}{n} \tag{6.5}$$

The coefficient $k(\alpha)$ can be found in Table 6.6:

Table 6.6 Coefficients for computing the Kolmogorov-Smirnov test statistic in a case of an exponential distribution with one parameter, based on Stephens (1974).

α	0.15	0.10	0.05	0.025	0.01
$k(\alpha)$	0.926	0.990	1.094	1.190	1.308

Finally, in the case of a Gumbel distribution, the CTGREF[19] (1978) used a simulation to compute a table of the Kolmogorov-Smirnov test statistic d. These values can be found in Table 6.7.

It should be noted that the CTGREF computed the experimental frequency using the following formula:

$$\hat{F}(x_{[r]}) = \frac{r - 0.3}{n + 0.4}$$

6.4 THE ANDERSON-DARLING TEST

Stephens showed (Ahmad *et al.*, 1988) that the Anderson-Darling test is one of the most powerful, in contrast to the relatively weak power of the Kolmogorov-Smirnov and chi-square tests.

6.4.1 Principle and Characteristics

The Anderson-Darling test involves a modification of its Kolmogorov-Smirnov counterpart which makes it possible to give more weight to the tails of the distribution. The interest of this test comes from the fact that it is able to reproduce the upper tail of the distribution in the

[19] The Centre technique du génie rural, des eaux et forêts, now known as CEMAGREF.

Table 6.7 The Kolmogorov-Smirnov test statistic for the calibration of a Gumbel distribution, based on CTGREF (1978).

n	$\alpha = 0.20$	$\alpha = 0.10$	$\alpha = 0.05$	$\alpha = 0.01$
8	0.21	0.25	0.28	0.33
10	0.19	0.23	0.25	0.30
12	0.18	0.21	0.23	0.28
14	0.17	0.19	0.22	0.26
16	0.16	0.18	0.20	0.25
18	0.15	0.17	0.19	0.23
20	0.14	0.17	0.19	0.22
25	0.13	0.15	0.17	0.20
30	0.12	0.14	0.15	0.18
35	0.11	0.13	0.14	0.17
40	0.10	0.12	0.13	0.16
45	0.10	0.11	0.13	0.15
50	0.09	0.11	0.12	0.15
60	0.09	0.10	0.11	0.13
80	0.08	0.09	0.10	0.12
100	0.07	0.08	0.09	0.11
> 100	$\frac{0.70}{\sqrt{n}}$	$\frac{0.80}{\sqrt{n}}$	$\frac{0.90}{\sqrt{n}}$	$\frac{1.10}{\sqrt{n}}$

context of an extrapolation. Unlike the Kolmogorov-Smirnov test, where the critical values are independent of the underlying distribution, the Anderson-Darling test is based on a specific distribution to compute the critical values. This means it has the advantage of being a more sensitive test, but it comes with the drawback that the critical values must be computed for each distribution. Currently, tables of critical values are available for the following statistical distributions: normal, lognormal, exponential, Weibull, generalized extreme values (GEV) and logistic (see Stephens, 1974; Ahmad *et al.*, 1988).

More precisely, the Anderson-Darling test consists of comparing the theoretical frequency $F_0(x)$ to the experimental frequency $\hat{F}(x)$ by calculating the statistic:

$$A^2 = \int_{-\infty}^{+\infty} \{F(x) - F_0(x)\}^2 \, w(x) dF(x) \tag{6.6}$$

where $w(x)$ is a weighting function.

The classical Anderson-Darling test corresponds to the following weighting function:

$$w(x) = \frac{1}{F_0(x)\{1 - F_0(x)\}} \tag{6.7}$$

which makes it possible to give more influence to the smallest and largest frequencies (while the chi-square test, for example, gives *de facto* a larger weight to the "median" frequencies). Figure 6.4 illustrates the general shape of the weighting function $w(x)$.

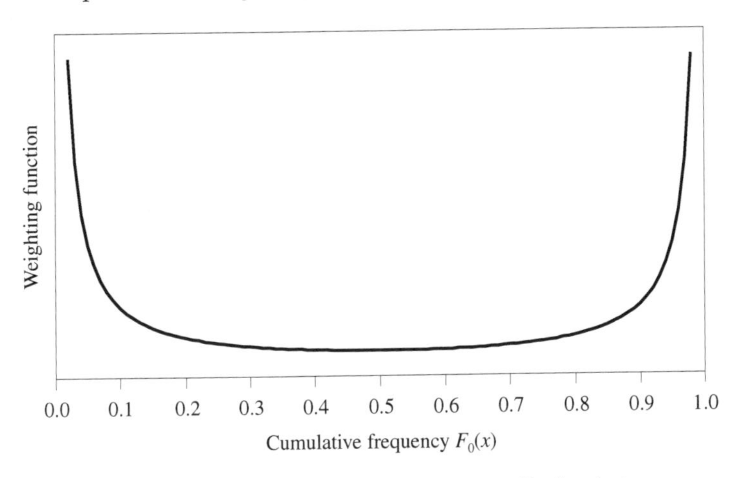

Fig. 6.4 Shape of the weighting function for the Anderson-Darling test.

6.4.2 Implementation

Once the series has been sorted in ascending order, the Anderson-Darling A^2 statistic (6.6), associated with the weighting function (6.7) is computed practically by discretizing the integral (6.6). This leads to the following formula:

$$A^2 = -n - \frac{1}{n}\sum_{i=1}^{n}(2i-1)\left[\ln F_0(x_{[i]}) + \ln\{1 - F_0(x_{[n+1-i]})\}\right] \tag{6.8}$$

In the case of a conformity test, Stephens (1974) computes the following critical values for A^2 when $n > 5$ (see Table 6.8):

Table 6.8 Anderson-Darling test statistic for $n > 5$, taken from Stephens (1974).

α	0.15	0.10	0.05	0.025	0.01
A^2	1.610	1.993	2.492	3.070	3.857

6.4.3 Used as a Goodness-of-Fit Test

For a normal distribution, Stephens (1974) developed the following formula for the test statistic:

$$A^2 = \frac{k(\alpha)}{1 + \frac{4}{n} - \frac{25}{n^2}} \tag{6.9}$$

where the factor $k(\alpha)$ is given in Table 6.9 for various common values of α.

Table 6.9 Coefficients for computing the Anderson-Darling test statistic for a normal distribution, based on Stephens (1974).

α	0.15	0.10	0.05	0.025	0.01
$k(\alpha)$	0.576	0.656	0.787	0.918	1.092

To test the goodness of fit of an exponential distribution with one parameter, Stephens developed the equation (6.10) for the test statistic:

$$A^2 = \frac{k(\alpha)}{1 + \frac{0.6}{n}} \tag{6.10}$$

where the factor $k(\alpha)$ can be found in Table 6.10:

Table 6.10 Coefficients for computing the Anderson-Darling test statistic for a 1-parameter exponential distribution, based on Stephens (1974).

α	0.15	0.10	0.05	0.025	0.01
$k(\alpha)$	0.992	1.078	1.341	1.606	1.957

To test the goodness-of-fit of a generalized extreme values distribution (GEV) with three parameters, the values of the A^2 statistic can be found in Ahmad *et al.* (1988) and are reproduced in Table 6.11.

Table 6.11 Critical values of the Anderson-Darling test statistic for a 3-parameter generalized extreme values (GEV) distribution, taken from Ahmad *et al.* (1988).

n	$\alpha = 0.15$	$\alpha = 0.10$	$\alpha = 0.05$	$\alpha = 0.01$
10	0.469	0.525	0.617	0.808
15	0.476	0.536	0.637	0.870
20	0.481	0.543	0.651	0.925
25	0.484	0.549	0.662	0.978
30	0.487	0.553	0.671	1.04
35	0.489	0.556	0.678	1.12
40	0.491	0.559	0.683	1.22
50	0.493	0.563	0.693	1.20
100	0.500	0.575	0.721	1.15

These values can be reached by using the following formula, also developed by Ahmad *et al.* (1988):

$$\alpha(A^2) \cong \sin^2\left(-1.128 + 0.5708A^2 - \frac{0.1867}{(A^2)^{\frac{3}{2}}} + \frac{0.8145}{A^2} - \frac{0.0737\ A^2}{\sqrt{n}} + \frac{0.1399}{A^2\sqrt{n}}\right) \tag{6.11}$$

6.4.4 Anderson-Darling Test Modified for Extreme Frequencies

By changing the weighting function (6.7) in:

$$w(x) = \frac{1}{1 - F_0(x)} \tag{6.12}$$

we obtain a test that is sensitive to the behavior of extreme frequencies (Ahmad *et al.*, 1988). This test procedure can be particularly useful when we are interested in extreme values, which in hydrology is usually the case.

In this case the test statistic, denoted AU^2 is computed as:

$$AU^2 = \frac{n}{2} - 2\sum_{i=1}^{n} F_0(x_i) - \sum_{i=1}^{n}\left(2 - \frac{i-1}{n}\right)\ln\{1 - F_0(x_i)\} \tag{6.13}$$

This modified version of the test also led Ahmad *et al.* (1988) to put forward a table of critical values for the GEV distribution, as well as the approximation formula given in equation (6.14):

$$\alpha(A^2) \cong \sin^2\left(-1.128 + 0.5708A^2 - \frac{0.1867}{(A^2)^{\frac{3}{2}}} + \frac{0.8145}{A^2} - \frac{0.0737\ A^2}{\sqrt{n}} + \frac{0.1399}{A^2\sqrt{n}}\right) \tag{6.14}$$

6.5 GOODNESS-OF-FIT INDICES

Another family of procedures involves measuring, in a standard frequency graph, the difference between an observed value x_i and the theoretical value $\hat{x}_i$ obtained from the calibrated model.

The procedure begins by defining the standardized difference d_i:

$$d_i = \frac{x_i - \hat{x}_i}{\bar{x}} \tag{6.15}$$

Then several *indices* can be defined, for example $\frac{1}{n}\sum d_i$, $\frac{1}{n}\sum d_i^2$, $\frac{1}{n}\sum |d_i|$, etc. For an example, the reader can refer to NERC (1975, page 150).

There are at least two criticisms that can be outlined regarding these indices:

- First of all, it requires applying an *empirical frequency formula* (see formula (5.22) for example), and hydrologists are far from being in agreement on these.
- Next, it gives no gradation of probability: usually these indices are used only for comparing performance of several frequency models. Thus, properly speaking, these procedures are not really goodness of fit tests.

6.6 COMPARISON OF MODELS

The Akaike Information Criterion (AIC) (Akaike, 1973) and Bayesian Information Criterion (BIC) (Schwarz, 1978) are the most frequently used criteria for selecting the best choice from all the possible models. There are also some other criteria such as the Minimum Description Length (MDL) (Rissanen, 1978) and the *Cp* of Mallows (1973). One model is better than another if it has a smaller AIC (or BIC). The AIC is based on the Kullback-Leibler distance in information theory, while the BIC is based on the integrated likelihood of Bayesian theory.

The Akaike Information Criterion (AIC) is computed as follows:

$$\text{AIC} = -2\log(L) + 2k \tag{6.16}$$

where L is the likelihood and k is the number of model parameters.

The Bayesian Information Criterion (BIC) seeks to select the model M_i that maximizes the *a posteriori* probability. This means it chooses the most likely model in view of the data.

This criterion is obtained as follows:

$$\text{BIC} = -2\log(L) + 2k\log(N) \tag{6.17}$$

where L is the likelihood, k is the number of parameters and N is the sample size.

These two criteria (BIC and AIC) make it possible to build a classification of statistical models while bearing in mind the principle of parsimony.

The AIC and BIC criteria have often been the object of empirical comparison (e.g. Bozdogan, 1987; Burnham and Anderson, 2002). In practice it has been observed that the BIC criterion selects models with smaller dimensions than the AIC, which is not surprising because the BIC penalizes more than the AIC for the number of model parameters (after $N > 7$).

If the complexity of the actual model does not increase with the size of the overall data, the BIC is the chosen criterion, otherwise the AIC is preferred (Burnham and Anderson, 2002).

6.7 CONCLUSIONS

For the internal check of the model, which is to say verifying its goodness of fit to the data, a first and systematic visual examination is recommended.

When it comes to testing goodness of fit, we advocate that practitioners avoid placing their confidence in the chi-square test, despite the fact that it is most often applied, by force of habit, because it has a notoriously weak power. It is better to replace it with the Anderson-Darling test modified for extreme frequencies (although it is worth noting that the statistic of this test has been developed only for the generalized extreme values distribution (GEV)). The AIC and BIC criteria are tools that make it possible to choose the most appropriate frequency model from the range of models.

CHAPTER 7

Uncertainties Analysis

At this stage of the analysis, after performing a number of steps, we have at our disposal our frequency model $\hat{F}(x)$. Now is the time to wonder about the *reliability* of the model, or the degree of confidence we can place in it.

From the engineer's viewpoint, the practical question to be solved is therefore: what is the design value x_D we should adopt.

7.1 SOURCES OF UNCERTAINTIES

Thirriot (1994) enumerated the *arbitraries* operating during the different steps of a frequency analysis:

- The first arbitrary is the choice of data to analyze. The inclusion of different periods, the inclusion or omission of a single value (if this happens to be an outlier), is all it takes to substantially modify the sensitive part of the distribution function in the extreme frequencies zone.
- The second arbitrary rests on the choice of formula for assessing the empirical frequency (see equation 5.22), required for undertaking certain specification techniques.
- The third source of uncertainty relates to the choice of the *frequency model* itself (see Chapter 4).
- The fourth arbitrary is the choice of the method for estimating the parameters, some of which, happily, spare us the effort of choosing a formula for assessing the empirical frequency.
- The fifth source of uncertainty results from using goodness of fit tests: some of these are like experts – they do not always give the same advice! So which test should we allow to guide our judgment?

Thirriot (1994) concluded this list as follows: *"Five arbitraries to arrive at an objective result! One must confess that the field is ripe for controversy. Fortunately, the results of the different paths are often not far apart with regards to the uncertainty governing the initial measurement or estimation. And then there are the schools of thought – or the habits - or conventions. Because we are well aware that, even if in many ways we are not absolutely certain that the result will be the best one, we have to do something and the imprecision will be minimal in comparison to the hazard. The 100-year return period flood could be next year or a thousand years away. So … ?"*

Miquel (1984, page 148), meanwhile, described three categories of uncertainties:

- *The information*: discrepancies in standard measurements (instantaneous discharge, discharge rating curves, rating curve shifts, etc.); discrepancies in the estimation discharges of extreme flood; lack of comprehensiveness.
- *The hypotheses*: discharge sample not completely natural (due to the presence of dams, canals, etc.); seasonal effects too marked; lack of stationarity, persistences, phenomena of large floods different from those of the common floods, etc..
- *The statistical model*: inadequate shape of the distribution; sampling uncertainties.

Miquel continues: *"This means it is important to assess, whenever possible, the uncertainties, and then incorporate the latter in the final estimation of the flood probabilities."* This author makes a distinction between "uncomputable" uncertainties and "computable" uncertainties.

Among the "uncomputable" uncertainties he cites certain progressive shifts in the discharge rating curve; lack of comprehensiveness regarding historical information; lack of stationarity over long periods; inadequacy of the chosen distribution $F(x)$.

The "computable" uncertainties include: error induced in the determination of discharge caused by errors in water depth measurements; uncertainties regarding observed extreme floods; the influence of artificial discharges; seasonal influence; the influence of non-stationarity over short periods of time, and finally, sampling uncertainties (which, from a practical viewpoint, are represented by a confidence interval).

Given this impressive inventory of uncertainties, the validation of an estimation is still not solved, but according to Miquel, it includes two steps:

- A *gamble*: this concerns the hypotheses and the "uncomputable" uncertainties. Only an analysis of sensitivity, if this approach can be carried out, can restore some confidence in the estimation.

- A *numerical estimation*: the confidence interval that sets the optimal quality of the estimation and the level of extrapolation possible, assuming that the other uncertainties are perfectly under control.

The main thing to remember from these general considerations is that the determination of the confidence interval, which is the subject of section 7.2 and which in practice, is the only uncertainty computable using the statistical theory of sampling (or at least the only "computable" uncertainty) is unfortunately not able to take into account all sources of uncertainties!

This means we need to:

- remain aware of our own ignorance and thus of the uncertainty that is inherent in frequency analysis;
- give priority to sensitivity studies of the various hypotheses, either quantified or unquantified.

7.2 THE CONFIDENCE INTERVAL

Uncertainty linked to the phenomenon of sampling fluctuation can be assessed using the classical *confidence interval* procedure. Constructing this interval is done by the so-called *standard error* method.

In this case, knowing the following three quantities is required:

- The estimation of the percentile, which, in the case of a Gumbel distribution is given by the expression:

$$\hat{x}_q = \hat{\alpha} + \hat{\beta}u_q \text{ with } u_q = -\log(-\log F)$$

- the standard deviation σ_{x_q}, determination of which is the subject of this section;
- the shape of the sampling distribution, to be discussed in the next subsection.

7.2.1 Shape of the Sampling Distribution

Most authors lean towards using a *normal distribution* for the sampling distributions of percentiles x_q. This habit can be explained by recognizing that:

- The normal distribution is the asymptotic form of a large number of distributions: this is the result in particular of the central limit theorem.
- Kite (1975) demonstrated that for the distributions most widely used in hydrology, the sampling distribution of a percentile is not significantly different from a normal distribution (although

it should be noted that Kite's conclusions result from chi-square and Kolmogorov-Smirnov goodness of fit tests and we are aware of the limitations resulting from the lack of power of these tests).

- It can be shown using information theory that, if the expectation of standard deviation is unknown, a normal distribution is the best possible choice (Harr, 1986).
- By applying a normal distribution to compute the confidence intervals even though another distribution could be better, one ends up with a mistake in the error computation: in this case it is a second-order error, which can probably be ignored.
- If the hypothesis of the normality of the sampling distribution is revealed to be erroneous for small values of the significance level α (corresponding to high levels of confidence $1 - \alpha$), this is still an acceptable choice for significance levels α that are not too low.

To these arguments we can add the *convenience* of using a normal distribution.

7.2.2 Standard Error of a Quantile, Method of Moments

In the case, for example, of a Gumbel distribution, we can undertake the following development:

Once the parameters α and β of the Gumbel distribution have been estimated using the method of moments, the expression of the percentile x_q can be written as follows (see equations 5.12 and 5.13):

$$\hat{x}_q = \hat{\mu} + 0.7797\hat{\sigma}\left(u_q - 0.5772\right) = \hat{\mu} + K_q\hat{\sigma} \quad (7.1)$$

where K_q is called the frequency factor in the now traditional formulation for a percentile in the USA (Chow, 1964):

$$K_q = 0.7797(u_q - 0.5772)$$

By using the expressions for computing the variance of a function of random variables, this becomes:

$$\mathrm{Var}\left(\hat{x}_q\right) = \mathrm{Var}\left(\hat{\mu} + K_q\hat{\sigma}\right) = \mathrm{Var}(\hat{\mu}) + K_q^2\mathrm{Var}(\hat{\sigma}) + 2K_q\mathrm{Cov}(\hat{\mu}, \hat{\sigma})$$

We also know (see for example NERC, 1975, page 101) that:

$$\mathrm{Var}(\hat{\mu}) = \frac{\sigma^2}{n}$$

$$\mathrm{Var}(\hat{\sigma}) = \frac{\mu_4 - \mu_2^2}{4\mu_2 n}$$

$$\mathrm{Cov}(\hat{\mu}, \hat{\sigma}) = \frac{\mu_3}{2\sigma n}$$

For a Gumbel distribution, the kurtosis coefficient β_2 is a known constant equal to 5.4. This means we can write $\mu_4 = 5.4\sigma_4$ and:

$$\mathrm{Var}(\hat{\sigma}) = \frac{5.4\sigma^4 - \sigma^4}{4\sigma^2 n} = \frac{4.4\sigma^4}{4\sigma^2 n} = \frac{1.1\sigma^2}{n}$$

The coefficient of symmetry γ_1 is also a constant, taking a value 1.1396. This gives $\mu_3 = 1.1396\sigma_3$ and:

$$\mathrm{Cov}(\hat{\mu}, \hat{\sigma}) = 0.5698\frac{\sigma^2}{n}$$

Based on these relations, the variance of the estimation of $\hat{x}_q$ can be written as:

$$\mathrm{Var}\left(\hat{x}_q\right) = \frac{\sigma^2}{n} + 1.1K_q^2\frac{\sigma^2}{n} + 2\cdot 0.5698K_q\frac{\sigma^2}{n};$$

or again:

$$\mathrm{Var}\left(\hat{x}_q\right) = \frac{\sigma^2}{n}\left(1.0 + 1.1396K_q + 1.1K_q^2\right)$$

Finally, by replacing σ^2 with its empirical estimation s^2 we have the following formula by *Dick and Darwin* (Gumbel, 1958, page 218):

$$\sigma_{x_q} = \frac{s}{\sqrt{n-1}}\sqrt{1.0 + 1.1396K_q + 1.1K_q^2} \tag{7.2}$$

or, by reintroducing the reduced Gumbel variable u_q, we obtain:

$$\sigma_{x_q} = \frac{s}{\sqrt{n-1}}\sqrt{0.709923 + 0.11657u_q + 0.668725u_q^2} \tag{7.3}$$

Table 7.1 presents a choice of values for the quantity under the square root, written as $\sqrt{D\ \&\ D}$ (for Dick and Darwin) for given return periods ranging from 1.5 to 2000 years:

Table 7.1 Values for computing the confidence interval according to Dick and Darwin (as a function of return period).

Return period T	1.50	1.58	2.0	2.33	5	10	15	20
$F(x_q)$	0.3333	0.3679	0.5000	0.5704	0.8000	0.9000	0.9333	0.9500
u_q	−0.0940	0.0000	0.3665	0.5772	1.4999	2.2504	2.6738	2.9702
$\sqrt{D\&D}$	0.8396	0.8426	0.9179	1.0000	1.5457	2.0878	2.4088	2.6364

Return period T	30	50	100	200	300	500	1000	2000
$F(x_q)$	0.9667	0.9800	0.9900	0.9950	0.9967	0.9980	0.9990	0.9995
u_q	3.3843	3.9019	4.6001	5.2958	5.7021	6.2136	6.9073	7.6006
$\sqrt{D\&D}$	2.9603	3.3684	3.9239	4.4813	4.8081	5.2204	5.7810	6.3428

Similar developments can be carried out for distributions other than the Gumbel. All these computations are relatively simple for 2-parameter distributions, but become far more complicated if we are dealing with 3 or more parameter distributions.

Example

Let us consider again the series of annual maximum mean daily discharges of the Massa (see Table 1.1 and section 5.2). The standard deviation $\hat{\sigma}$ is 11.75; the sample size $n = 80$. The two estimators of the parameters of the Gumbel distribution obtained using the method of moments are $\hat{\beta} = 9.16$ and $\hat{\alpha} = 68.17$.

We would like to compute the 80% confidence interval of the quantile corresponding to a 100-years return period. The standard deviation is given by the formula of Dick and Darwin as follows:

$$\sigma_{x_{0.99}} = \frac{11.75}{\sqrt{79}} \cdot 3.9239 = 5.19$$

The quantile corresponding to a 100-year return period is obtained from Table 7.1 as:

$$\hat{x}_{0.99} = \hat{a} + \hat{b}u_{0.99} = 68.17 + 9.16 \times 4.6 = 110.3$$

Then we apply the classical equation for a confidence interval built using the standard error method:

$$\Pr\left(\hat{x}_q - z_{1-\frac{\alpha}{2}}\sigma_{x_q} < X_q < \hat{x}_q + z_{1-\frac{\alpha}{2}}\sigma_{x_q}\right) = 1 - \alpha$$

In a table of the standard normal distribution, we look for the 90% percentile $z_{0.90} = 1.28 \Rightarrow$

$$\Pr(110.3 - 1.28 \times 5.19 < X_{100} < 110.3 + 1.28 \times 5.19) = 0.80$$

$$\Pr(103.7 < X_{100} < 116.9) = 0.80$$

7.2.3 Standard Error of a Quantile, the Maximum Likelihood Method

When the parameters of the distribution have been estimated with the maximum likelihood method, the procedure for computing the standard error of a quantile is different (see for example Kite, 1988 or Masson, 1983). The variance of the estimation is obtained using the general relation for a quantile $x_q = f(\alpha, \beta, \ldots; q)$. For a Gumbel distribution, the function has only two unknown parameters: as $x_q = f(\alpha, \beta; q)$. As q is constant, we can apply the Taylor linearization method, which leads to:

$$\text{Var}(x_q) \cong \left(\frac{\partial x}{\partial \alpha}\right)^2 \text{Var}(\alpha) + \left(\frac{\partial x}{\partial \beta}\right)^2 \text{Var}(\beta) + 2\frac{\partial x}{\partial \alpha}\frac{\partial x}{\partial \beta}\text{Cov}(\alpha, \beta) \tag{7.4}$$

The partial derivatives can be obtained in this case by using the analytical equation of a quantile. We know in addition from mathematical statistics that the variances-covariances matrix of the parameters is the inverse of the *Fisher information matrix* obtained from the derivatives of the likelihood function *V*:

$$\begin{pmatrix} \mathrm{Var}(\alpha) & \mathrm{Cov}(\alpha,\beta) \\ \mathrm{Cov}(\alpha,\beta) & \mathrm{Var}(\beta) \end{pmatrix} = \begin{pmatrix} -\dfrac{\partial^2 V}{\partial \alpha^2} & -\dfrac{\partial^2 V}{\partial \alpha \partial \beta} \\ -\dfrac{\partial^2 V}{\partial \alpha \partial \beta} & -\dfrac{\partial^2 V}{\partial \beta^2} \end{pmatrix}^{-1} \quad (7.5)$$

For a Gumbel distribution, after computations we find (Kite, 1988; Masson, 1983):

$$\sigma_{x_q} = \frac{\hat{\beta}}{\sqrt{n}} \sqrt{1.1086 + 0.514u_q + 0.6979u_q^2} \quad (7.6)$$

Example

We will use the same example as the previous one, but apply the maximum likelihood method to estimate the parameters of the Gumbel distribution. In this case we obtain $\hat{\beta} = 10.05$ and $\hat{\alpha} = 67.91$.

For $T = 100$ years, $u_q = 4.6$. The quantity under the square root of (7.6) thus takes a value of 18.24 and the standard deviation will be:

$$\sigma_{x_{100}} = \frac{10.05}{\sqrt{80}} \sqrt{18.24} = 4.80$$

It can be seen that the standard error computed in the case of an estimation done with the maximum likelihood method is smaller than the one obtained using the relation (7.3): $4.80 < 5.14$. This is conform to the following fact highlighted by mathematical statistics: the estimators with maximum likelihood always show minimal variance, even if they are sometimes biased.

In this case the 80%-confidence interval is given by:

$$\Pr(110.3 - 1.28 \times 4.37 < X_{100} < 110.3 + 1.28 \times 4.37) = 0.80$$

$$\Pr(104.7 < X_{100} < 115.9) = 0.80$$

7.2.4 Another Expression of Standard Error

The two approaches described above are based on approximations. This means they can be refined. CTGREF[20] (1978), for example, advocates the construction of a confidence interval using the following equation for the Gumbel distribution:

[20] Now CEMAGREF.

$$T_{1,2} = \frac{\frac{z_{1-\alpha/2}}{\sqrt{n}}\sqrt{1.0 + 1.1396K_q + 1.1K_q^2} \pm \frac{z_{1-\alpha/2}^2}{n}(1.1K_q + 0.5772)}{1 - \frac{1.1z_{1-\alpha/2}^2}{n}} \tag{7.7}$$

where $z_{1-\alpha/2}$ is the quantile of the standard normal variable corresponding to a chosen level of $(1 - \alpha)$ for the confidence interval and K_q is the frequency factor for the Gumbel distribution corresponding to the chosen return period:

$$K_q = 0.7797(u_q - 0.5772)$$

Thus the confidence interval is defined by:

$$\mathrm{P}\left(\hat{x}_q - T_2 s < X_q < \hat{x}_q + T_1 s\right) = 1 - \alpha \tag{7.8}$$

It can be seen that this confidence interval is not symmetric and that, if we ignore the second term and the denominator of equation (7.7), we obtain again the Dick and Darwin equation.

Example

Considering the same example as above, we find for a 80%-confidence interval:

$$\Pr(110.3 - 0.49 \times 11.75 < X_{100} < 110.3 + 0.66 \times 11.75) = 0.80$$

$$\Pr(104.5 < X_{100} < 118.1) = 0.80$$

Note that the interval obtained this way is not symmetric.

7.3 CONFIDENCE INTERVAL AND RETURN PERIOD

7.3.1 The "Classical" Approach

A frequency analysis includes the selection and preparation of data, the choice of a frequency model, and the estimation of the model parameters, which leads to the specification of $\hat{F}(x)$.

Given this, a quantile x_q can be assessed by using the relation corresponding to the frequency model that has been adopted (for example: $\hat{x}_q = \hat{\alpha} + \hat{\beta}u_q$ for a Gumbel distribution), for a chosen cumulative probability q. The exceedance probability of p of $\hat{x}_q$ is given by $1 - q$.

It is usual to associate this quantile $\hat{x}_q$ with a *confidence interval*, characterized by the *standard error* $\sigma_{\hat{x}_q}$.

Most often we use this information to "get an idea" of the confidence we can have in the result.

Sometimes the *design value* x_D to adopt is determined based on the standard error using an equation such as:

$$x_D = \hat{x}_q + C\sigma_{\hat{x}_q} \tag{7.9}$$

where C is a factor, commonly known as the *confidence factor*, dependent on the shape of the sampling distribution and on the desired level of confidence 1 – α (cf. Fig. 7.1).

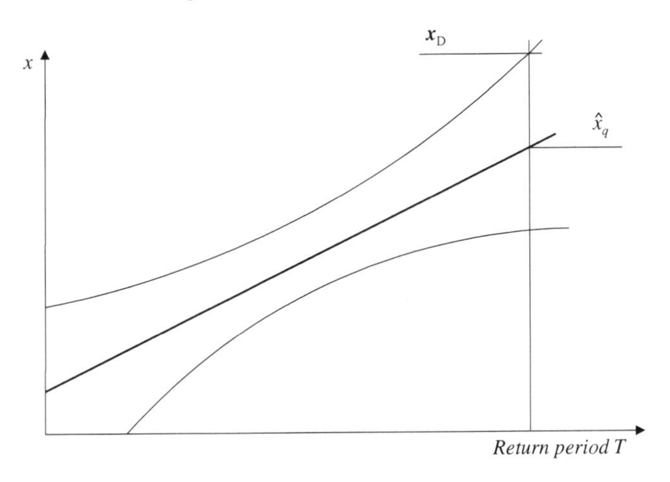

Fig. 7.1 $(1 - \alpha)^{th}$-confidence interval and the traditional approach for determining the design value.

It is exactly the choice of this confidence level that constitutes a central aspect of the question expressed by Bernier and Veron (1964): *"In practice, we would use a 70% confidence interval, because if the 95% interval corresponds to a higher probability of concealing the true value of* x_q, *it usually has an amplitude too large to be of practical significance"*. This comment is pause for thought and hints at the existence of an unresolved problem.

In Poland it is common to use a 68% confidence interval (Sokolov *et al.*, 1976). In the ex-Soviet Union (*same reference)*, the project value is determined by a relation of this type:

$$x_D = \hat{x}_q \left(1.0 + \frac{a\varepsilon_P}{\sqrt{n}}\right)$$

where a is a coefficient based on the available degree of information (ranging from 0.7 to 1.5), ε_P is "the standard deviation of the frequency curve" (defined as a function of the chosen exceedance probability p and the coefficient of variation (Sokolov *et al.*, 1976, page 26) and n is the sample size.

In Switzerland, OFEG[21] (2001) recommends to use the upper limit of the 80% confidence interval as the design value.

The problem of choosing the confidence level that should be adopted for determining the design value has not been solved satisfactorily using this approach, considered as classical because it is often used.

7.3.2 Return Period Revisited

More careful examination shows that we are faced with two distinct aspects of the exceedance probability or the failure probability:

- The exceedance probability $\hat{p} = 1 - \hat{F}(x)$, usually characterized by its inverse value: the return period $\hat{T} = 1/\hat{p}$, which qualifies the hazard.
- The exceedance probability of the quantile $\hat{x}_q$, characterized by the sampling distribution of this quantile, which is to say by it confidence interval, which is representative of *statistical uncertainty*.

These two exceedance probabilities must be combined in order to obtain a single probability that corresponds to the *risk that the decision-maker or the project owner decides to assume*: the failure probability.

It is this ***failure probability*** of the structure being designed that is of central interest, which is the exceedance probability p:

$$p = \Pr(X > x_D) \tag{7.10}$$

It is in fact, in accordance with its definition, the inverse of the return period $T(x)$, but a return period that corresponds to the actual distribution $F(x)$, which is unknown, and not to the estimated return period $\hat{T}(x)$, used implicitly to compute the quantile $\hat{x}_q$. We will see later on that these two return periods are not equal, which means that unfortunately, $T(x) \neq \hat{T}(x)$! $\hat{T}(x)$ is a biased estimator of $T(x)$.

7.3.3 Approach by Integrating the Sampling Distributions of the Quantiles

A first approach to this problem consists of estimating the "actual" return period $T(x)$ by integrating the exceedance probabilities of a given value x over the domain of the probabilities $F(x)$. For any quantile $\hat{x}_q$, the sampling distribution makes it possible to write the exceedance probability as:

$$\Pr\left(X_q \geq x_D\right) = 1 - F_{Xq}(x_D) \tag{7.11}$$

[21] Office Fédéral des Eaux et de la Géologie, now the Office Fédéral de l'Environnement (OFEV).

where $F_{Xq}(x)$ is the distribution function (sampling distribution) of the quantile x_q. Then we can integrate this probability for all possible values of $F(x)$ or q (see Fig. 7.2). It becomes:

$$p = \Pr(X > x_D) = \int_0^1 \{1 - F_{Xq}(x_D)\} dq \tag{7.12}$$

Finally, the "actual" return period $T(x)$ will be given by $1/p$.

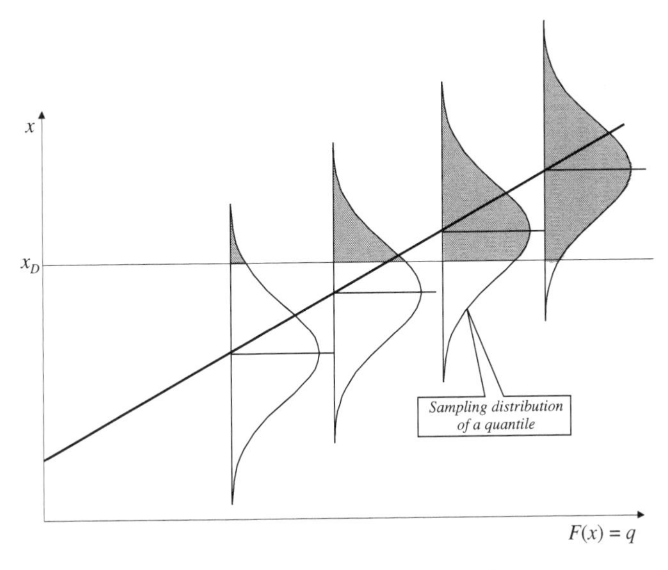

Fig. 7.2 Principle of computing the exceedance probability based on the sampling distributions of the quantiles x_q.

Application to a Normal Distribution

A normal distribution is used first to evaluate this approach. Chowdhury and Stedinger (Stedinger *et al.* 1993, pages 18-30) developed the formula in order to compute the applicable sampling distribution (for $n \geq 15$ and $1 - \alpha/2 \leq 0.95$ where $1 - \alpha$ defines the confidence interval of the quantile):

$$\bar{x} + \zeta_{\alpha/2,q} S_X < x_q < \bar{x} + \zeta_{1-\alpha/2,q} S_X \tag{7.13}$$

where

$$\zeta_{\alpha,q} \cong \frac{z_q + z_\alpha \sqrt{\dfrac{1}{n} + \dfrac{z_q^2}{2(n-1)} - \dfrac{z_\alpha^2}{2n(n-1)}}}{1 - \dfrac{z_\alpha^2}{2(n-1)}} \tag{7.14}$$

For reasons of convenience, the numerical computation is carried out by choosing for X a standard normal distribution (with a mean of $\mu = 0$ and a variance of $\sigma^2 = 1$), without loss of generality. The integration over the F (or q) domain is carried out numerically using the trapezoidal approach, with steps of 0.001, bearing in mind that the first term (for $q = 0$) is equal to 0 and the last (for $q = 1$) is equal to 1.

We denote the "supposed" return period $\hat{T}$, or the one serving to define the quantile $\hat{x}_q$, and the "actual" return period T computed by the procedure we are about to describe. The following results were obtained for a sample size $n = 20$ (see Table 7.2):

Table 7.2 Results of a computation by integration of the confidence intervals.

$\hat{T}$	2	5	10	25	50	100	200	500	1000
T	2	4.81	9.08	20.08	35.34	60.24	100.0	181.8	277.8

The actual return period T is thus systematically smaller than the estimated return period $\hat{T}$ or T^*. These results are shown graphically in Figure 7.3.

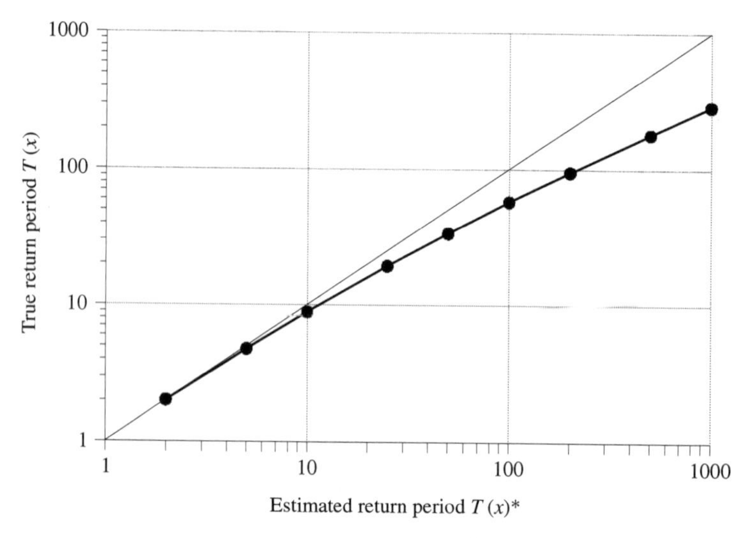

Fig. 7.3 Relation between the "estimated" return period $\hat{T}$ or T^* and the "actual" return period T for a normal distributions and a sample size of n=20.

7.3.4 Back to the Confidence Interval

For comparison purposes, it would seem useful here to compute the "classical confidence interval" to which this approach corresponds using the relation (7.12), still for the same example, which is for a normal distribution calibrated by the method of moments for a sample size $n = 20$. The result of this computation is given in Table 7.3.

Table 7.3 "Classical" confidence interval resulting from application of equation (7.12).

$\hat{T}$	2	5	10	25	50	100	200	500	1000
T	2	4.81	9.08	20.08	35.34	60.24	100.0	181.8	277.8
$1-\alpha$ (%)	0	11	18	27	33	38	43	49	52

The two first lines of this table are identical to those of Table 7.2. The last line indicates the confidence interval to use, based on the calibration performed with the sample size $n = 20$, to obtain a design value x_D compatible with the approach based on the relation (7.12).

This table leads to the following two assessments:

- the confidence interval varies with the return period;
- the order of magnitude of the obtained confidence intervals is clearly lower than those that are commonly adopted in practice (compared to the 68% advocated by Polish hydrologists or the 70% from Bernier and Veron, 1964).

In light of these considerations, formula (7.12) appears to be a valuable tool that deserves to be applied more widely.

7.3.5 Beard's Approach

Beard (1960) published a different approach to the same problem. Basically, he performed a hydrological application of the Proschan formula (1953) which gives the expectation of the exceedance probability of a quantile x_q for a normal distribution:

$$E\{\Pr(X > \hat{x} + \hat{\sigma} z_q)\} = \Pr\left(t_{n-1} > z_q\sqrt{\frac{n}{n+1}}\right) \tag{7.15}$$

where $\hat{x}$ and $\hat{\sigma}$ are the two parameters of the normal distribution estimated using the method of moments, z_q is the standard normal variable corresponding to the cumulative frequency q, t_{n-1} is the Student random variable with $n - 1$ degrees of freedom and n is the sample size. This makes it easy to compute the values of T as a function of $\hat{T}$, as previously done. These values are reported in Table 7.4:

Table 7.4 Result of the computation according to Beard (1960).

$\hat{T}$	2	5	10	25	50	100	200	500	1000
T	2	4.75	8.85	19.27	33.62	57.16	94.61	178.9	286.7

Figure 7.4 compares Beard's approach (1960) to the integration method described in equation (7.12).

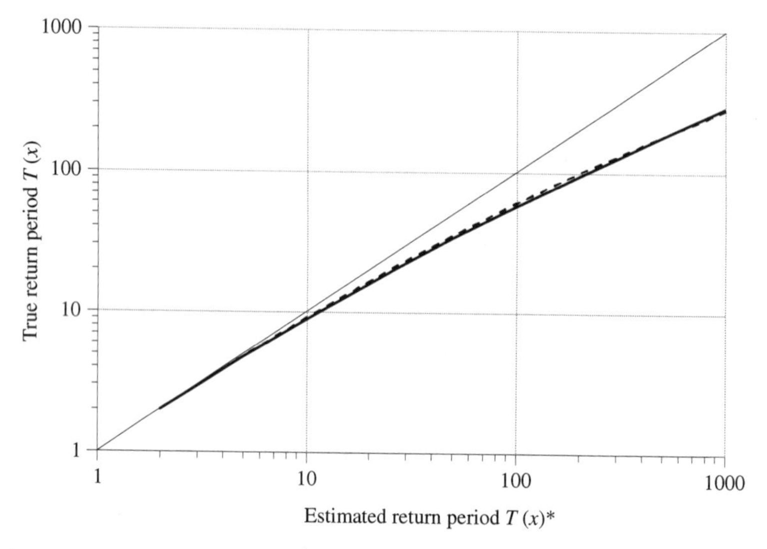

Fig. 7.4 Relation between T and T^* obtained using the Proschan formula (Beard, 1960) indicated by a solid line, compared to the integration approach using formula (7.12), dotted line.

The high degree of concordance of these two results allows us to conclude that Beard's approach, which considers the expectation of the exceedance probability, is equivalent to the approach that involves the integration of the sampling distributions.

It would be interesting to explain this concordance statistically or to show that it could be only a coincidence. It should also be mentioned that Beard's method does not yield unanimity (Lang, 1996).

7.3.6 Simulation Approach

Beard's procedure described above can only be applied for a normal population. Gunasekara and Cunnane (1991) proposed and tested a procedure involving simulation that frees us from this type of distribution. The steps are as follows:

- With the available sample of size n, we calibrate the distribution $F_{\hat{\theta}}(x)$.
- Then we randomly generate from this distribution $F_{\hat{\theta}}(x)$ N random samples of size n.
- For each sample, we carry out a new calibration of distribution $F_{i,\hat{\theta}}(x)$ from which we compute the desired quantiles. Finally we

determine the exceedance probability p_i of this quantile $x_{i,q}$ for the distribution $F_{\hat{\theta}}(x)$.

- The sought after exceedance probability p is the mean of the N values of p_i obtained.

7.4 CONFIDENCE INTERVAL USING THE PARAMETRIC BOOTSTRAP

Yet another way to build the confidence interval of a quantile is to use a statistical technique such as the bootstrap (Davison and Hinkley, 1997). Assuming that we have a sample $(x_1, ..., x_i, ..., x_n)$ of size n. We calibrate a distribution F and estimate its parameters using a method M. Thus we obtain from this an estimation $\hat{\theta}$ of a vector of parameters θ. The idea of the bootstrap is to resample based on the distribution $F_{\hat{\theta}}(x)$. This procedure involves the following steps:

1. Generate R samples of size n (R being sufficiently large such as R = 10 000).
2. Estimate the quantile of interest $\hat{x}_q$ for each sample.
3. This gives a sample of quantiles $q = \{\hat{x}_{q,1}, \cdots, \hat{x}_{q,R}\}$.

Next, for the sorted sample, we determine the empirical quantiles $x_{q \cdot \alpha/2}$ and $x_{q,1-\alpha/2}$ of the nonexceedance probability of $\alpha/2$ and $1 - \alpha/2$ respectively. This gives a confidence interval of x_q at level $(1 - \alpha) : \left[x_{q,\alpha/2}, x_{q,1-\alpha/2}\right]$.

It should be noted that this method is more precise than an asymptotic computation when n is small.

Example

Again we use the annual maximum mean daily discharge of the Massa at Blatten. Figure 7.5 compares the 80% confidence levels obtained using the Dick and Darwin relation and applying a parametric bootstrap. The excellent concordance resulting from these two approaches is to be highlighted.

7.5 CONCLUSIONS

In conclusion we have to say once again that the theory of frequency analysis, as it now stands, is not able to provide the practitioner with clear guidance. Nonetheless, the following elements should be remembered:

- computing the confidence interval, even if it is estimated, is necessary in order to appreciate, even subjectively, the sampling uncertainty;

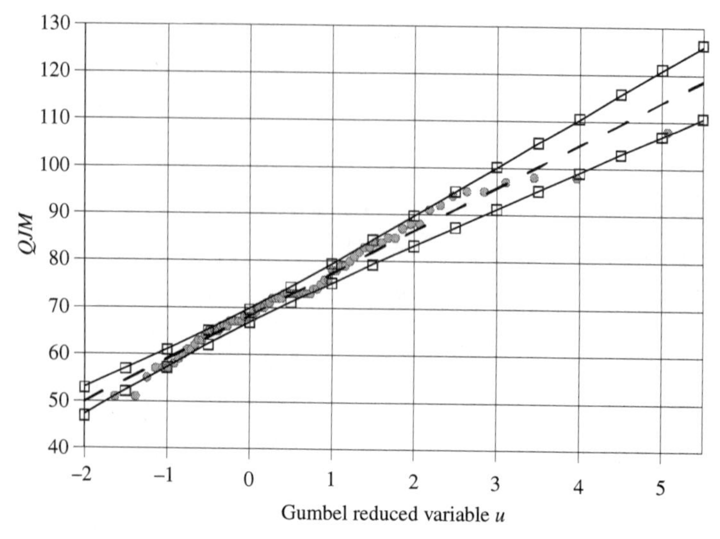

Fig. 7.5 Annual maximum mean daily discharge of the Massa at Blatten. 80% confidence interval using the parametric bootstrap (solid lines) and according to Dick and Darwin (black squares).

- the sampling uncertainly is far from being the only uncertainty that should be taken into consideration;
- hazard (exceedance probability $p = 1/T$) and sampling uncertainty (characterized by the confidence interval of $\hat{x}_q$) can be combined to give a single probability of failure: this aggregation resulting from equation (7.12) should probably be subjected to some simulation techniques in order to reduce the inherent errors due to the approximation formula of confidence intervals. Such a technique (using simulation) would also make it possible to take into account, simultaneously the added uncertainties induced by the "annualization" of a POT series (see Chapter 8).

CHAPTER 8

Using the Frequency Model

Using a frequency model consists in part of using the relation that provides the value of a quantile x_q as a function of the cumulative frequency $q = F(x_q)$ or, conversely, of computing the cumulative frequency (and thus the return period) corresponding to a given value x. This aspect, which may seem self-evident, was already addressed several times, especially in Chapters 1 and 2, and will be described only briefly here.

In addition we can compute the *probability of occurrence* of a particular hydrological event. These computations generally operate on the assumption that the adopted frequency model refers to *annual probabilities* or that the observation periods (the statistical trials) are in *years*.

However there can be cases in which the analysis does not deal with annual periods, especially when we are trying to take into account as much of the available data as possible, with the ongoing aim of comprehensiveness and homogeneity. This is the case, for example, when we are performing an analysis of POT series or series of k greatest annual events ("inflated" series). In such instances we are dealing with *infra-annual* series and it is advisable, before computing the probability of occurrence, to transform the infra-annual series to annual series – an operation we will call *annualization.*

Next we will look at an example of applying some frequency models in order to reach a more general synthesis: IDF curves (intensity - duration - frequency) which summarize in a single model a series of frequency models that each corresponds to a different rainfall duration.

We will examine next a method that solves the composition of errors problem by exposing a little known approach: the Rosenblueth method.

Finally, in Section 8.6, we will propose an alternative approach, the Poissonian counting method, that allows us to find/provide an answer for return periods equal to or less than one year, and also to handle time series that do not present at least one event per year.

8.1 THE PROBABILITY OF HYDROLOGICAL EVENTS

8.1.1 A Simple Application of a Frequency Model

The simplest application of a frequency model consists of determining the quantile $x(F)$ where F is the cumulative frequency corresponding to the desired return period (see for example equation (1.8)).

Example

In Chapter 5 (Section 5.2) we have estimated the parameters of a Gumbel distribution for the maximum annual mean daily discharges of the Massa River, looking for the maximum daily mean discharge corresponding to a 100 year return period. The mathematical expression of the Gumbel distribution is as follows:

$$F(x) = \exp\left\{-\exp\left(-\frac{x-\alpha}{\beta}\right)\right\} \text{ and } x(F) = \alpha + \beta u(F) \text{ where } u(F) = -\ln(-\ln F)$$

with $\hat{\alpha} = 68.17$ and $\hat{\beta} = 9.16$.

The discharge we are looking for is obtained as follows:

$$F = 1 - \frac{1}{T} = 1 - \frac{1}{100} = 0.99 \text{ ; } u_{0.99} = -\ln\{-\ln(0.99)\} = 4.6 \text{ ;}$$

$$x(0.99) = q_{100} = 68.17 + 9.16 \times 4.6 = 110.3 \text{ [m}^3\text{/s]}$$

Conversely, we could look for the return period of a given event $\{X>x\}$.

Example

What is the return period of a discharge of 90 [m^3/s]? Using the reduced Gumbel variable u we obtain:

$$u = \frac{90 - 68.17}{9.16} = 2.38 \text{ ; } F = \exp\{-\exp(-2.38)\} = 0.912 \text{ ;}$$

$$T = \frac{1}{1-F} = \frac{1}{1-0.912} = 11.4 \text{ years}$$

8.1.2 The Probability of Hydrological Events

The techniques applied for determining the probability of events were discussed in detail in Chapter 2 (Section 2.2). Hereafter we will present one example of how this type of computation can be carried out.

Example

A project on the Massa River at Blatten will take two years to be completed. The plan is to secure the worksite by means of a cofferdam and a temporary diversion of the river. What is the mean daily discharge[22] that we need to consider if we accept a 5% probability that the worksite will be submerged during the 2-year construction period?

We will start from relation (2.5), which we can rewrite as follows (see equation (2.6)):

$$T = \frac{1}{1 - \{1 - \Pr(W \geq 1)\}^{\frac{1}{N}}}$$

$$T = \frac{1}{1 - (1 - 0.05)^{\frac{1}{2}}} = 39.5 \text{ years}$$

So we need to consider the mean daily discharge corresponding to a 39.5 year return period $x(0.9747) = q_{39.5} = 68.17 + 9.16 \times 3.6642 = 101.7$ [m^3/s].

8.2 ANNUALIZATION OF INFRA-ANNUAL SERIES

In hydrology, and generally speaking in all fields connected to climate, the length of the series under consideration (or in other words the sample size) is often limited. Thus in order to use, as much as possible, all the information available, two approaches are used (see also Fig. 3.9):

- A sample is built with all the independent events higher than a given threshold. The resulting series is called a peak over a threshold or POT series (see Section 3.2.3)[23]. This type of approach, compatible with the "renewal processes" techniques used especially in France (Miquel, 1984) requires several developments in order to end up ultimately with the annual probabilities.
- A series is composed with the k greatest annual events (more commonly called an "inflated" series). This method requires a transformation to annual probabilities, which is simpler than the preceding approach.

Note that "inflated" series are often considered as a solution that is less than satisfying (Michel, 1989) and that it would be better to use a POT series.

---

[22] The replacement of the mean discharge with the peak discharge, which is more judicious in this case, will be discussed in Section 8.4.

[23] As already mentioned in Section 3.2, the choice of a POT series is also motivated by a concern for homogeneity.

Finally, it is worth remembering that, even if the statistical reliability of the results could be improved, the *climatological reliability* is only linked to the duration of the observations.

8.2.1 Series of *k* Greatest Annual Values ("Inflated" Series)

This type of series is obtained by selecting the k greatest events in each observation period of the complete series of independent events. The observation period is usually defined as one year.

For this series, it is possible to calibrate a distribution function written as $F_G(x)$ (where G designates the "inflated" series). The problem consists then of determining the "annualized" distribution function $F_A(x)$.

One might think that it suffices to consider that the observation periods are no longer "years" but are instead "$1/k$ year"—for example a trimester if the value of k is equal to 4. From this perspective the computation would be straightforward: one would only need to work with periods of $1/k$ years. However proceeding this way makes no sense: by construction, the series indeed involves k events per year but there is no reason that these events are regularly spaced so that one occurs in each $1/k$ year! This means that a different approach to the problem is needed.

Let us consider $F_A(x)$, the probability that the maximum *annual* value is less than the value x. If the maximum annual value is lower than x, it obviously follows that the same must be true for the k greatest values in the year. By considering the k annual maximum values as independent events, we can apply the composition of probability as follows:

$$F_A(x) = \underbrace{F_G(x) \cdot F_G(x) \cdot \ldots \cdot F_G(x)}_{k \text{ times}}$$

or:

$$F_A(x) = \{F_G(x)\}^k \tag{8.1}$$

By inverting equation (8.1), we obtain:

$$F_G(x) = \{F_A(x)\}^{\frac{1}{k}} \tag{8.2}$$

Example with the Gumbel Distribution

When the Gumbel distribution is applied to model a series of the k greatest annual values, written using the reduced variable u_G as:

$F_G(x) = e^{-e^{-u_G}}$, where the index G refers to the "inflated" series.

the corresponding "annual" distribution (denoted with index A) is likewise a Gumbel distribution written as:

$$F_A(x) = e^{-e^{-u_A}} \tag{8.3}$$

The relations (8.1) and (8.3) above make it possible to write:

$$u_A = -\ln\{-\ln F_A(x)\} = -\ln\left[-\ln\{F_G(x)\}^k\right]$$

$$u_A = -\ln\{-k \ln F_G(x)\} = -\ln k - \ln\{-\ln F_G(x)\}$$

The relation between the reduced Gumbel variables corresponding to each of these two distributions is thus written as follows:

$$u_A = u_G - \ln k \text{ and } u_G = u_A + \ln k$$

The straight lines representing the relation between the reduced Gumbel variable and the quantiles of the variable of interest are therefore parallel: they have the same slope which is β (Fig. 8.1). The position parameters α are linked by the following relations:

$$\alpha_A = \alpha_G + \beta \ln k \text{ and } \alpha_G = \alpha_A - \beta \ln k$$

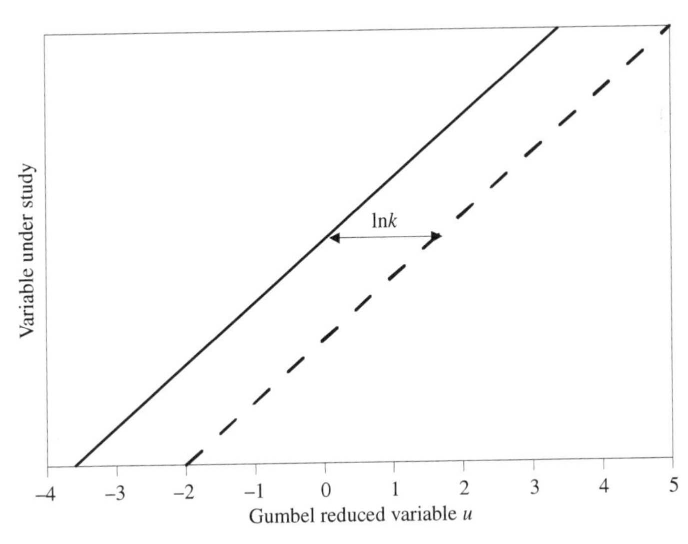

Fig. 8.1 Plots of an "inflated" series (dotted line) and an annualized distribution (solid line).

8.2.2 Peak Over a Threshold Series

A POT series is obtained by selecting, from a complete series of independent events, all the values higher than a given threshold x_0. Let us denote the size of the resulting series by n. It is then possible to calibrate to this series a probability distribution written as $F_T(x)$, where the index T denotes the POT series. The problem now consists of determining the distribution function $F_A(x)$ corresponding to the *annual* distribution of probability.

It is generally agreed that the occurrence of a given hydrological event (for example, $Y = \{X > x_0\}$) during a fixed period of time can be described by a Poisson process:

$$\Pr(Y = k) = e^{-\lambda} \frac{\lambda^k}{k!} \tag{8.4}$$

that expresses the probability that k events will occur (we define in this case an event as an occurrence of $\{X > x_0\}$ during the interval of time Δt, usually one year). The Poisson distribution has only one parameter λ, which represents the mean number of events that take place during the time period Δt:

$$\lambda = \frac{n}{n_A} \tag{8.5}$$

where n_A is the number of periods (years) under study.

However there exist also other possible distributions besides the Poisson (discussed later in this chapter).

The distribution function $F_A(x)$ gives the probability that an event is lower or equal to a given value x (which, however, by construction is higher than the fixed threshold x_0). The probability that k independent events are all lower than x can then be computed by applying the multiplication rule for independent events:

$$\Pr(Y_1 < x, Y_2 < x, \cdots, Y_k < x) = \Pr(Y_1 < x)\Pr(Y_2 < x)\cdots\Pr(Y_k < x) = \{F_T(x)\}^k \tag{8.6}$$

In addition, the probability of occurrence of k events greater than the fixed threshold x_0 is given by the Poisson distribution. Consequently:

$$F_A(x) = \sum_{k=0}^{\infty} \Pr(Y = k)\, \{F_T(x)\}^k \tag{8.7}$$

where $\Pr(Y = k)$ is the probability of occurrence of k events $\{X > x_0\}$ during the selected time period Δt (one year, for example) and $F_A(x)$ is the probability that no event greater than the threshold x_0 will occur.

By replacing $\Pr(Y = k)$ with its expression (8.4) it becomes:

$$F_A(x) = \sum_{k=0}^{\infty} \frac{1}{k!} e^{-\lambda} \lambda^k F_T(x)^k = e^{-\lambda} \sum_{k=0}^{\infty} \frac{1}{k!} \{\lambda F_T(x)\}^k \tag{8.8}$$

The sum contained in the above equation is none other than a series expansion of an exponential:

$$e^z = \sum_{k=0}^{\infty} \frac{1}{k!} z^k$$

so that the relation under study can be easily simplified as:

$$F_A(x) = \exp[-\lambda\{1 - F_T(x)\}] \tag{8.9}$$

or conversely:

$$F_T(x) = 1 + \frac{1}{\lambda} \ln F_A(x) \tag{8.10}$$

The two last equations ((8.9) and (8.10)) are known as the *Langbein-Takeuchi relations.*

Remarks

- These relations are equivalent to the formula proposed by Langbein (1949), and revisited by Takeuchi (1984).
- In the case where the Poisson distribution is not appropriate to describe the occurrence of events probability, Takeuchi (1984) proved that these relations remain valid as long as the concerned return period is larger than two or three years, which is practically always the case in hydrological applications.
- We often find the following approximation for large F_T: $-\lambda(1-F_T) \to 0$ and as $e^x \underset{x \to 0}{\approx} 1 + x$ we have $F_A \approx 1 - \lambda(1 - F_T)$. The return period can thus be expressed as $T_A = \frac{1}{1 - F_A} = \frac{1}{\lambda(1 - F_T)}$.

The Case of Exponential and Gumbel Distributions

POT series are often modeled using a 2-parameter exponential distribution (denoted with the indice expo) as:

$$F_T(x) = 1 - e^{-\frac{x-\alpha_{\text{expo}}}{\beta_{\text{expo}}}}$$

By using equation (8.9), after algebraic simplification we obtain:

$$F_A(x) = e^{-\lambda e^{-\frac{x-\alpha_{\text{expo}}}{\beta_{\text{expo}}}}}$$

Then we can express the opposite of the exponent as:

$$\lambda e^{-\frac{x-\alpha_{\text{expo}}}{\beta_{\text{expo}}}} = e^{\ln\lambda} e^{-\frac{x-\alpha_{\text{expo}}}{\beta_{\text{expo}}}}$$

and thus obtain for $F_A(x)$:

$$F_A(x) = e^{-e^{-\left(\frac{x-\alpha_{\text{expo}}}{\beta_{\text{expo}}} - \ln\lambda\right)}}$$

which is the expression of a Gumbel distribution (denoted with the indice gum) with:

$$\frac{x - \alpha_{\text{gum}}}{\beta_{\text{gum}}} = \frac{x - \alpha_{\text{expo}}}{\beta_{\text{expo}}} - \ln\lambda$$

By rewriting this relation as:

$$\frac{x - \alpha_{\text{gum}}}{\beta_{\text{gum}}} = \frac{x - \alpha_{\text{expo}} - \beta_{\text{expo}} \ln\lambda}{\beta_{\text{expo}}}$$

we can deduce the relation between the parameters of the exponential distribution (expo) calibrated for the POT series and the parameters of the Gumbel distribution (gum), the annualized distribution as follows:

$$\left.\begin{aligned} \beta_{gum} &= \beta_{\text{expo}} \\ \alpha_{gum} &= \alpha_{\text{expo}} + \beta_{\text{expo}} \ln\lambda \end{aligned}\right\} \tag{8.11}$$

Thus the two distributions have the same asymptotic slope (same gradex).

Example

Let us look again at the data presented in Section 6.1, the peak discharges at Nozon, recorded over a period of 8.25 years between 1923 and 1931. The sample size is 60 in this case. These discharges were calibrated using an exponential distribution with parameters $\alpha_{\text{expo}} = 2.02$ and $\beta_{\text{expo}} = 1.32$. The annualization to a Gumbel distribution can be expressed as follows using equation (8.11)

$$\lambda = \frac{60}{8.25} = 7.27 \ ; \ \beta_{\text{gum}} = 1.32 \ ; \ \alpha_{\text{gum}} = 2.02 + 1.32 \ln(7.27) = 4.64$$

The relation between the reduced Gumbel variable and the quantity of interest (in fact, the peak discharge of the event), for the POT series and for the annualization, is represented in Figure 8.2.

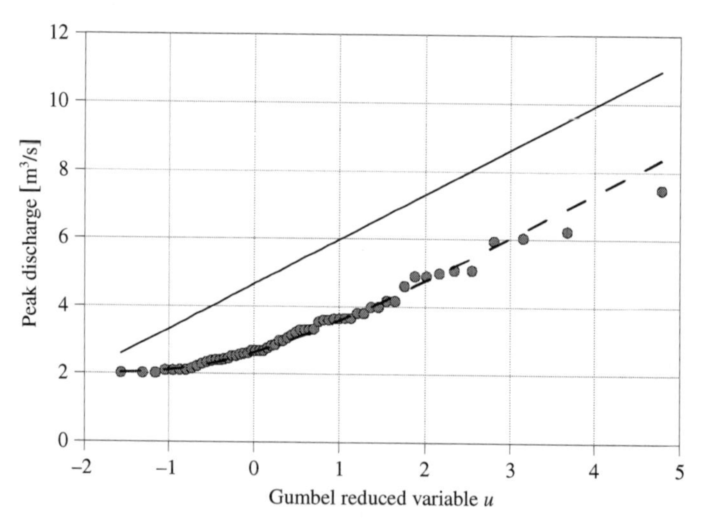

Fig. 8.2 Relation between the reduced Gumbel variable and the peak discharge for the POT series (dotted line) and for the annualized distribution (solid line).

The case of a generalized Pareto distribution (GPD) and of a generalized extreme values (GEV) distribution

In a case where a POT series is modeled using a generalized Pareto distribution (denoted with the indice gpd):

$$F_{gpd}(x) = 1 - \left(1 - \gamma_{gpd}\frac{x - \alpha_{gpd}}{\beta_{gpd}}\right)^{\frac{1}{\gamma_{gpd}}}$$

Stedinger *et al.* (1993) proved that the corresponding annual distribution is the generalized extreme values (GEV):

$$F_{gev}(x) = \exp\left\{-\left(1 - \gamma_{gev}\,\frac{x - \alpha_{gev}}{\beta_{gev}}\right)^{\frac{1}{\gamma_{gev}}}\right\}$$

with the following relations between the parameters:

$$\left.\begin{aligned} \gamma_{gev} &= \gamma_{gpd} = \gamma \\ \beta_{gev} &= \beta_{gpd}\,\lambda^{-\gamma} \\ \alpha_{gev} &= \alpha_{gpd} + \beta_{gpd}\,(1 - \lambda^{-\gamma})/\gamma \end{aligned}\right\} \tag{8.12}$$

8.2.3 Summary and Conclusions

It emerges from the above relations that application of the POT series approach in connection with "annual" Gumbel (GUM) distributions and generalized extreme values distributions (GEV) is justifiable and sound. In essence, the distributions adapted to the corresponding POT series are, respectively, the exponential distribution (EXP) and the generalized Pareto distribution (GPD).

It should be noted that once the threshold x_0 has been chosen, the exponential distribution has only a single undetermined parameter while the generalized Pareto distribution has no more than two.

Admittedly there remains some additional uncertainty regarding the parameter λ of the Poisson distribution. However experience has shown that once the mean number is larger than 1.65, the technique works as well as the one based on annual series (NERC, 1975); see also the boxed text in section 3.2.3.

Table 8.1 presents a summary of the annualization formulas corresponding to POT series, as well as the relation between the distribution parameters for a POT series and for "annual" series.

Table 8.1 Summary of annualization formulas.

<table>
<tr><th>Distribution for POT series</th><th>"annual" distribution</th></tr>
<tr><td>Generalized Pareto distribution (GPD):
$F(x) = 1 - \left\{1 - \gamma \frac{(x - \alpha)}{\beta}\right\}^{\frac{1}{\gamma}}$
$x = \frac{\beta}{\gamma}\left[1 - \{1 - F(x)\}^{\gamma}\right] + \alpha$
if $\gamma = 0 \Rightarrow$ exponential distribution</td><td>Jenkinson distribution (GEV):
$F(x) = \exp\left[-\left\{1 - \gamma \frac{(x - \alpha)}{\beta}\right\}^{\frac{1}{\gamma}}\right]$
$x = \frac{\beta}{\gamma}\left[1 - \{- \ln F(x)\}^{\gamma}\right] + \alpha$
if $\gamma = 0 \Rightarrow$ Gumbel distribution</td></tr>
<tr><td colspan="2">GPD → GEV: $\gamma_{gev} = \gamma_{gpd} = \gamma\ ;\ \beta_{gev} = \beta_{gpd}\lambda^{-\gamma};$
$\lambda = \frac{n}{n_a}\ \alpha_{gev} = \alpha_{gpd} + \frac{\beta_{gpd}(1 - \lambda^{-\gamma})}{\gamma}$</td></tr>
<tr><td>Exponential (EXPO):
$F(x) = 1 - \exp\left(-\frac{x - \alpha}{\beta}\right)$
$x = \alpha - \beta\ln\{1 - F(x)\}$</td><td>Gumbel (GUM):
$F(x) = \exp\left\{-\exp\left(-\frac{x - \alpha}{\beta}\right)\right\}$
$x = \alpha - \beta\ \{-\ln F(x)\}$</td></tr>
<tr><td colspan="2">EXPO → GUM: $\beta_{gum} = \beta_{expo} = \beta\ ;$
$\lambda = \frac{n}{n_a}\ ;\ \alpha_{gum} = \alpha_{expo} + \beta\ln\lambda$</td></tr>
<tr><td colspan="2">Langbein-Takeuchi relationship:
$F_A(x) = \exp[-\lambda\{1 - F_T(x)\}]$</td></tr>
</table>

8.3 THE FREQUENCY USE OF NONSYSTEMATIC DATA

Using historical data in frequency analysis is not particularly recent. A special issue of the *Journal of Hydrology* in 1987 (no. 96) was devoted to the analysis of extreme hydrological events. In this issue, more than 20 of the articles were devoted to the use of historical floods for the purpose of estimating flood probabilities. O'Connell *et al.* (2002) used paleohydrological information in a Bayesian context. Parent and Bernier (2003) developed the POT model to take into account historic data. They applied their approach to the Garonne River and showed that the introduction of 12 historical data reduced the 90% confidence level by 40% for a discharge corresponding to a 1000-year return period. Payrastre (2005) did a feasibility study concerning the application of historical data to study extreme floods in small rivers.

8.4 FREQUENCY USE OF IDF CURVES

This section is intended to show that the results of a frequency analysis as described up to this point can be applied to develop more general models of practical utility, such as intensity - duration – frequency (IDF) curves.

8.4.1 What is an IDF Curve?

An IDF curve explains and summarizes the behavior of a universally observed rainfall distribution, or in other words:

"For the same frequency of exceedance – and therefore the same return period – the intensity of a rain event is stronger when its duration is shorter" or the corollary "for rains of equal duration, a precipitation will be more intense as its frequency of exceedance decreases (thus its return period increases)." (Musy and Higy, 2011, p. 129).

It is well known that rainfall *intensity* (the "I" in the acronym IDF) is highly variable. To lay people this variability may seem totally random. However, to those paying a little more attention, even if they are not meteorologists, it appears that there are certain observable "rules" governing the phenomenon: the chaos is by no means total!

From observation we can deduce that, "on average", the more the *duration* under study (the "D" in IDF) increases the more the intensity diminishes: the shortest rainstorms are often the most violent, while rainfalls of longer duration, such as autumnal rains, are generally quite moderate (low intensity).

Finally, to better characterize the phenomenon we associate each event with a *probability of exceedance.* In practice, engineers and hydrologists talk in terms of *return period*: such an event occurs, on average, over a long period of time, every x years (see Section 2.1). Here we are talking about the *frequency of occurrence* (the "F" in IDF).

This means that observations of precipitations and statistical analysis of the results can be used to find a relation of the behavior that links *intensity, duration* and *frequency*: the IDF curve.

Frequency information about rains can be synthesized in the form of *frequency models* that give, for a rainfall of a chosen duration d, the value of a quantile $x(F)$. Then we can proceed to a series of analyses of the extreme rainfalls in each of the durations d of interest.

It is equally possible to present a different relation between rain, return period and duration. This consists of representing, for each return period T selected, the rain as a function of duration. This presentation is called an *IDF curve.* Its advantage is that it is well suited to interpolation for different durations d.

As a general rule, when working with IDF curves rainfall is expressed in terms of *intensities* in [mm/h] rather than in volume [mm]. The choice of intensity to characterize the rainfall allows us to represent a large number of curves simultaneously, each curve corresponding to a given return period (Fig. 8.3).

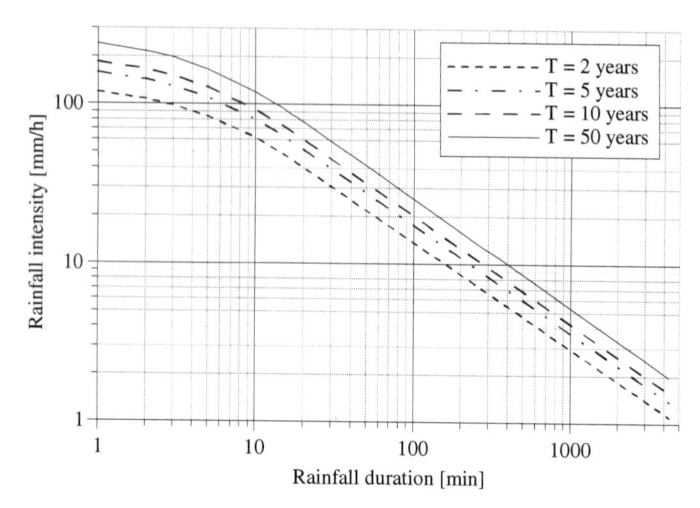

Fig. 8.3 Example of regional IDF curves for the Canton of Geneva (Switzerland).

In the literature a great number of curve shapes based on empirical relations can be found. Here we will present only two that are commonly used in Switzerland and France, and a third that combines them.

8.4.2 The Talbot Relationship

The Talbot relationship for a given return period T expresses the rainfall intensity I as a function of the duration d of the rainstorm as follows:

$$I(d) = \frac{c}{e + d} \tag{8.13}$$

where c and e are the two parameters of this equation (the reason for choosing c and e as the parameters in this case will be given below). This relation is recommended in the Swiss Association of Road and Transportation Experts standards[24] (VSS, 2000), although with different nomenclature and units.

[24] Called *"normes des professionnels suisses de la route"* in French.

8.4.3 The Montana Relationship

For a given return period T this relationship is expressed as:

$$I(d) = ad^{-b} \tag{8.14}$$

where a and b are the two parameters to be estimated.

On a double logarithmic scale graph, the Montana relationship is represented by a straight line as follows:

$$\ln I(d) = \ln a - b \ln d$$

8.4.4 Combined Talbot-Montana Relationship

Several authors including Hörler (1977) and Grandjean (1988) have proposed a combined model that involves adopting the Talbot relationship for shorter durations (of less than about 20 minutes) and the Montana relationship for longer durations. Figure 8.4 illustrates this approach.

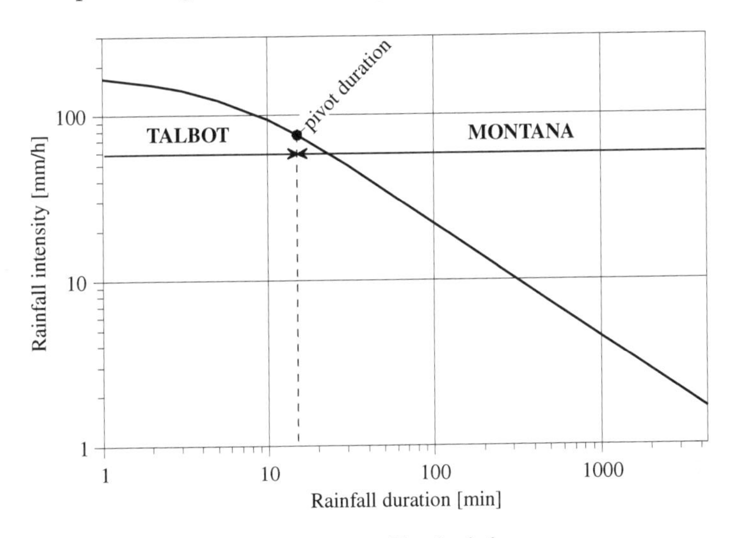

Fig. 8.4 Principle of the Talbot-Montana combined relation.

To define this "combined" relationship, we first introduce a limit of validity called the *pivot duration* d_P.

To formalize this new relation, we first have to set the same value for the pivot duration to ensure continuity:

$$\frac{c}{e + d_P} = ad_P^{-b} \Rightarrow c = (e + d_P)ad_P^{-b} \tag{8.15}$$

Consequently, from the five parameters (a, b, c, e and d_P) we can remove c.

The second constraint is the continuity of the slope at the pivot duration. Thus we should equal the derivatives of the intensity $I(d)$ with respect to the duration d.

For the Talbot relationship this becomes:

$$\frac{\mathrm{d}}{\mathrm{d}d}\left(\frac{c}{e+d}\right)=\frac{(e+d)c'-c(e+d)'}{(e+d)^2}=-\frac{c}{(e+d)^2} \tag{8.16}$$

For a Montana relationship this equality becomes:

$$\frac{\mathrm{d}}{\mathrm{d}d}\left(ad^{-b}\right)=a\{-bd^{-(b+1)}\} \tag{8.17}$$

By equating these two derivatives in the pivot duration and replacing c with its expression (equation (8.15)) we have:

$$abd_P^{-b}d_P^{-1}=\frac{(e+d_P)ad_P^{-b}}{(e+d_P)^2}\Rightarrow bd_P^{-1}=\frac{1}{e+d_P} \tag{8.18}$$

Finally we obtain the expression for the parameter e and simplify the expression of c:

$$e=\frac{d_P}{b}-d_P \tag{8.19}$$

$$c=\frac{a}{b}\,d_P^{1-b} \tag{8.20}$$

Thus the combined Talbot-Montana relationship requires only three parameters: a, b and d_P.

8.4.5 A Global Model

The "simple combined" model we just looked at has to be developed for each return period T being analyzed.

Assuming that all the *subjacent* models are of the same type, it is also possible to develop a *global* model (Meylan *et al.*, 2005) that gives the intensity I as a function of the duration d but also of the return period T.

Let us suppose that all the subjacent frequency models applied for the analysis of rainfalls of duration d are, as an example, Gumbel distributions, which we can write as:

$$x_q=\alpha+\beta u_q$$

where x_q is the quantile under study, α and β are the two parameters of the Gumbel distribution and u_q is the Gumbel reduced variable corresponding to the quantile q.

Having repeated the analysis for different durations d we now have the values $\alpha(d)$ and $\beta(d)$. Experience has shown that if the Talbot-Montana model applies to intensities $I(d)$ it is equally suited to parameters $\alpha(d)$ and $\beta(d)$. Thus we can formulate the following global model:

$$I(d,T) = \alpha(d) + \beta(d)u(T) \tag{8.21}$$

where

- $\alpha(d)$ is the function expressing the variation of the Gumbel parameter α (the mode) as a function of the duration;
- $\beta(d)$ is the relation expressing the variation of the Gumbel parameter β as a function of the duration;
- $u(T)$ is the reduced Gumbel variable, expressed as a function of the return period by:

$$u(T) = -\ln\left\{-\ln\left(1 - \frac{1}{T}\right)\right\} \tag{8.22}$$

If the shape of the IDF curve adopted to describe $\alpha(d)$ and $\beta(d)$ is the Talbot-Montana one (as in the combined model) we obtain a model with six parameters (three for each parameter α and β of the Gumbel distribution).

Table 8.2 shows the parameters of a global IDF model as it applies to the territory of the Canton of Geneva (Meylan *et al.*, 2005).

For the expression of a general model that corresponds to various distributions used to represent the frequency of rainfalls of duration d, the reader is referred to Koutsoyiannis (1998).

Table 8.2 Parameters of a global IDF model for the Canton of Geneva (Switzerland).

Gumbel parameter	TALBOT			MONTANA	
	c	*e*	*Pivot*	*a*	*b*
α	1039.06	8.78	17.50	266.06	0.666
β	289.13	7.46	17.50	86.13	0.701

Example

Our aim is to calculate the intensity of the 25-year return period of a rainstorm with a duration of 20 minutes.

The parameter α is expressed as: $\alpha = 266.1 \times 20^{-0.666} = 36.19$.

β can be written: $\beta = 86.1 \times 20^{-0.701} = 10.54$.

The reduced Gumbel variable is equal to $u = -\ln\{-\ln(1-1/25)\} = 3.2$.

And we find $i = 36.19 + 10.54 \times 3.2 = 69.9$ [mm/h].

8.5 CONSIDERING UNCERTAINTIES IN THE CASE OF A FUNCTION WITH SEVERAL VARIABLES

In the practice of frequency analysis we are often looking for a quantity that is a function of several variables, each of them associated by an error characterized, for example, by its standard error.

In such a case it is useful to combine several random variables. We will examine three possible approaches for handling this, the first two very superficially (these two approaches having been abundantly described in general textbooks on statistics) so that we can focus on a technique that is still not widely known but which is extremely handy—the Rosenblueth technique. This method is often the only one applicable in the case of complex functions of random variables, especially when they have been obtained through simulation.

Example

In the case of the calibration of the annual maximum mean daily discharges of the Massa at Blatten, which we have often used as an example, it becomes possible to obtain (for example) the discharge corresponding to a 100-year return period, as well as its confidence interval or its standard error.

If we are interested in the peak discharge rather than the mean daily discharge, we can use a relation such as:

$$Q_p = Q_{mj} C_p$$

where Q_p is the peak discharge, Q_{mj} is the mean daily discharge and C_p is an experimental coefficient: the *peak coefficient*. The two quantities Q_{mj} and C_p are associated with their own standard error.

8.5.1 Exact Relations

In a limited number of cases, we can use *exact* relations derived from a rigorous mathematical development. Here we will illustrate the simplest case where Y is the random variable obtained from a linear combination of the random variables X_i, $i = 1,\ldots,n$.

$$Y = \sum_{i=1}^{n} a_i X_i + b, \text{ then:}$$

$$\mu_Y = \sum_{i=1}^{n} a_i \mu_{X_i} + b \tag{8.23}$$

$$\sigma_Y^2 = \sum_{i=1}^{n} a_i^2 \, \sigma_{X_i}^2 + \sum_{\substack{i=1 \\ i \neq j}}^{n} \sum_{j=1}^{n} a_i a_j \mathrm{Cov}(X_i, X_j) \tag{8.24}$$

8.5.2 A Linearization Approach

In more complex cases where *exact* relations do not apply, we have recourse to an *approximation* formula obtained with a Taylor expansion. As an example let us consider (Pugachev, 1982) the function:

$$Y = \varphi(X_1, \ldots, X_n)$$

Then, the first-order limited expansion leads to:

$$\mu_Y \cong \varphi(\mu_{X_1}, \cdots, \mu_{X_n}) + \frac{1}{2}\sum_{i=1}^{n}\sum_{j=1}^{n}\left(\frac{\partial^2\varphi}{\partial x_i \partial x_j}\right)\sigma_{X_iX_j} \tag{8.25}$$

$$\sigma_Y^2 \cong \sum_{i=1}^{n}\sum_{j=1}^{n}\left(\frac{\partial^2\varphi}{\partial x_i \partial x_j}\right)\sigma_{X_iX_j} \tag{8.26}$$

8.5.3 The Rosenblueth Method

In order to avoid the sometimes tiresome computations involved in Taylor's linearization technique (in fact at times they might be impossible – we are thinking particularly about results coming from simulation models) Rosenblueth (1975) proposed an estimation method based on a *discretization* of the density functions of random variables.

The basic principle of the approach consists of replacing the continuous distribution of a random variable with an "equivalent" discrete distribution. Figure 8.5 illustrates the principles of Rosenblueth's approach.

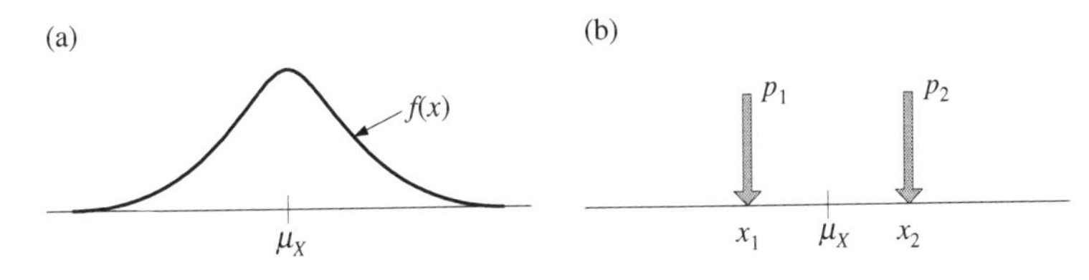

Fig. 8.5 Principle of the Rosenblueth method: a) continuous distribution, b) "equivalent" discrete distribution.

The discrete distribution is written as:

$$p(x) = \sum_{i=1}^{m} p_i \partial(x - x_i)$$

where m is the number of points of the discretization; x_i denotes the position of point i, p_i is the probability concentrated at point x_i, and ∂ is the Dirac operator.

In this context the term "equivalent" signifies that the *moments* of the two distributions (continuous and discrete) must be equal. Depending on the case, we can restrict ourselves to ensuring this equality for the j first moments of the distribution, $j = 2, 3, 4, \ldots$ The relations below provide an example of this required equivalence for the two first moments:

1. $\sum_{i=1}^{m} p_i = 1$, results from the axiom $\Pr(\Omega) = 1$;

2. $\sum_{i=1}^{m} p_i x_i = \mu$, first-order moment;

3. $\sum_{i=1}^{m} p_i (x_i - \mu)^2 = \sigma^2$, second-order centered moment.

Two-point Symmetric Distribution

To illustrate this method, let us consider the simplest case consisting of a symmetric distribution and an equivalent two-point discrete distribution ($m = 2$). By using conditions 1) and 2) above we obtain:

$$\left. \begin{array}{l} p_1 = \frac{1}{2}; \; p_2 = \frac{1}{2} \\ x_1 = \mu - \sigma; \; x_2 = \mu + \sigma \end{array} \right\} \tag{8.27}$$

With these relations the statistical moments are computed using their definition (a reminder):

$$E[Y^k] = \sum_{i=1}^{m} p_i y_i^k$$

The variance is obtained with the classical equation:

$$\sigma^2 = E(X^2) - \{E(X)\}^2$$

An example of the application of this procedure is given below.

Other Cases

Apart from this first case (two-point symmetric distribution) the Rosenblueth method can be applied using approximations with three or more points. It can even take into account the case of dependent random variables (correlated). An interesting extension was proposed by Harr (1989) in order to simplify the computation considerably when the number of variables to be combined increases.

Example

Again using the series of annual maximum mean daily discharges of the Massa River at Blatten, we want to obtain an estimation of the *peak discharge* corresponding to a return period of 200 years, with an 80% confidence interval.

We already have the estimation of the annual maximum mean daily discharge corresponding to a return period of 100 years and its standard deviation (see the example in Section 7.2.2):

$$\hat{x}_{0.99} = 110.3 \; ; \; \sigma_{x_{0.99}} = 5.14$$

To obtain the peak discharge we examine the $Q_{mj} - Q_p$ pairs of the mean daily discharges and the peak discharges for the samples available, which are from 1992 to 2003. From these data pairs we retain only the 33 pairs where $Q_{mj} > 50$ [m^3/s] and we compute the peak coefficient defined by:

$$C_p = \frac{Q_p}{Q_{mj}}$$

The sample mean of the peak coefficient is $\overline{C}_p = 1.23$ while its standard error is $s_{\overline{C}_p} = 0.0174$.

We begin by using the *exact* relation for the product of two independent random variables $Z = XY$ (see e.g. Ventsel, 1973):

$$\mu_Z = \mu_X \mu_Y$$

$$\sigma_Z^2 = \sigma_X^2 \sigma_Y^2 + \mu_X^2 \sigma_Y^2 + \mu_Y^2 \sigma_X^2$$

this becomes:

$$Q_p = 1.23 \times 110.3 = 135.7$$

$$s_{Q_p}^2 = 0.0174^2 \times 5.14^2 + 1.23^2 \times 5.14^2 + 110.3^2 \times 0.0174^2 = 43.66; \; s_{Q_p} = 6.61$$

To illustrate the method, Table 8.3 shows the development of the computation according to Rosenblueth for a two-point distribution.

Table 8.3 Illustration of the Rosenblueth method (case of the Massa River, two-point distribution).

j	P_{ij}	QJM_i	CP_j	QP_{ij}	$P_{ij}QP_{ij}$	$P_{ij}(QP_{ij})^2$
1	0.25	106.16	1.21	128.77	32.19	4'145.56
1	0.25	115.44	1.21	140.03	35.01	4'902.01
2	0.25	106.16	1.25	132.38	33.10	4'381.22
2	0.25	115.44	1.25	143.95	35.99	5'180.67
					136.28	18 609.45

Thus we obtain $Q_p = 136.3$ and $s_{Q_p} = \sqrt{18\,609.45 - 136.3^2} = \sqrt{36.13} = 6.01$.

This approximation is completely satisfactory and the peak discharge can be matched with an 80% confidence interval by using a normal approximation:

$$\Pr(136.3 - 1.28 \times 6 < Q_P < 136.3 + 1.28 \times 6) = 80\%$$

$$\Pr(129 < Q_P < 144) = 80\%$$

8.6 FREQUENCY ANALYSIS BY COUNTING

Originally, the goal of frequency analysis was to model events of large amplitudes (rare or extreme) by means of statistical distributions called *extreme value distributions.* Current growing concerns relating mostly to the environment led to an interest in analyzing more frequent events. In this context, it is natural to wonder about events corresponding to one year or even shorter return periods. These questions are usually related to the "discharge" variable, especially linked with issues of water quality or management of biotopes.

However this question raises a major issue: How can we use, for a discharge with a 6-month return period for example, the formula that gives us the cumulative frequency:

$$F = 1 - \frac{1}{T} = 1 - \frac{1}{0.5} = -1\ !$$

Applying this formula would clearly be a violation of one of the fundamental axioms concerning the computation of probabilities!

Another category of problems raises another question: What to do when the series of observations does not contain at least one event per year? This can happen, for example, when we are studying the behavior of hydraulic systems equipped with storm overflow systems. It is entirely possible that in a certain number of years there will not be a single overflow.

We can solve this problem by considering the observed events not as the realization of *Bernoulli trials* (a year being considered as a statistical trial), but as the realization of a *Poisson process* (Meylan, 2010).

8.6.1 The Poisson Process

Let us consider the number of events $N(t)$ that occur in the interval $[0, t]$.

The stochastic process $\{N(t);\ t \geq 0\}$ is called a ***point process*** (Ross, 2010; Solaiman, 2006).

A ***Poisson process*** is a point process that verifies the following three conditions:

- The process $N(t)$ is *homogeneous in time.* This means that the probability of observing k events in an interval of a given duration t depends only on t and not on the position of the interval on a temporal axis.
- The increasing *increments* of the process $N(t)$ is done at independent and stationary growth. This means that for every

system of separate intervals, the numbers of events that happen within these intervals are independent variables. The process is said to have *no memory*.

- The probability that two or more events will occur within the time interval Δt is negligible.

8.6.2 Poissonian Counting

Counting

The number of points is equal to the rank order in a series of the greatest events sorted in *decreasing order*, which we denote as i so as not to confuse it with the rank order r of a series sorted in increasing order, which is used in classical frequency analysis. So here we will work with a POT series rather than a series of annual events.

If N is the sample size we have the following relation:

$$k_i = i = N + 1 - r \tag{8.28}$$

Counting Rate

The classical statistical relation applied for the counting rate can be written as:

$$\lambda_i = \frac{k_i}{n_A}$$

where n_A is the number of years of observations.

However in order to ensure coherence with the empirical frequency formulas used in classical frequency analysis, we can use instead (see equation 5.22 and Table 5.1):

$$\lambda'_i = \frac{k_i - c}{n_A + 1 - 2c} \tag{8.29}$$

where c is a coefficient taking values between 0 and 0.5.

Using $c = 0.5$ (Hazen plotting position) this formula becomes:

$$\lambda' = \frac{k - 0.5}{n_A} \tag{8.29b}$$

If this correction is applied, leading to an estimation denoted here with a prime (" ' "), we can now compute again the *corrected count* k'_i as:

$$k'_i = \lambda'_i n_A \tag{8.30}$$

when $c = 0.5$ (HAZEN) : $k'_i = k_i - 0.5$ (8.30b)

This new count k'_i will be used to compute the confidence interval.

Return Period

At this point the return period is obtained as the inverse of the counting rate (or λ_i if it was not corrected using relation 8.29):

$$T_i = 1/\lambda_i \text{ or } T_i' = 1/\lambda_i'$$

Confidence Interval

If K is a Poisson random variable (a counting) the bilateral confidence interval of k at level $(1 - \alpha)$ is given by (CEA, 1978, p. 80):

$$\Pr\left\{\frac{1}{2}\,\chi^2_{2k}\left(\frac{\alpha}{2}\right) \le k \le \frac{1}{2}\,\chi^2_{2(k+1)}\left(1 - \frac{\alpha}{2}\right)\right\} = 1 - \alpha \qquad (8.31)$$

where $2k$ and $2(k + 1)$ are the number of degrees of freedom to be used, usually denoted as ν and χ^2 is the chi-square distribution, function of the cumulative frequency of interest (here $\alpha/2$ and $(1 - \alpha/2)$).

The bounds of the bilateral confidence interval at level $(1 - \alpha)$ can be expressed as:

$$k_{\text{inf}} = \frac{1}{2}\,\chi^2_{2k}\left(\frac{\alpha}{2}\right) \text{ and } k_{\text{sup}} = \frac{1}{2}\,\chi^2_{2(k+1)}\left(1 - \frac{\alpha}{2}\right) \qquad (8.31b)$$

It is worth noting that the confidence interval here concerns the *return period* and not, as is usually the case, the variable of interest x.

8.6.3 Parametric Model

The relations described above make it possible to attribute, to each value in the POT series, a return period and even a confidence interval for this return period.

In this section we intend to model the observations using an *equation* (so here we are not requiring a statistical distribution anymore!).

The possible equations for such an application are, in light of the considerations induced by the distribution of *excesses* (see Section 4.2.2), the exponential equation and the generalized Pareto equation.

Modeling Using the Exponential Equation

We use the following formulation:

$$x = a + b\ln T$$

For a POT series, it is possible to define the return period T for each term in the series. For example, by applying Hazen's correction, it becomes:

$$T = \frac{n_A}{i - 0.5} \qquad (8.32)$$

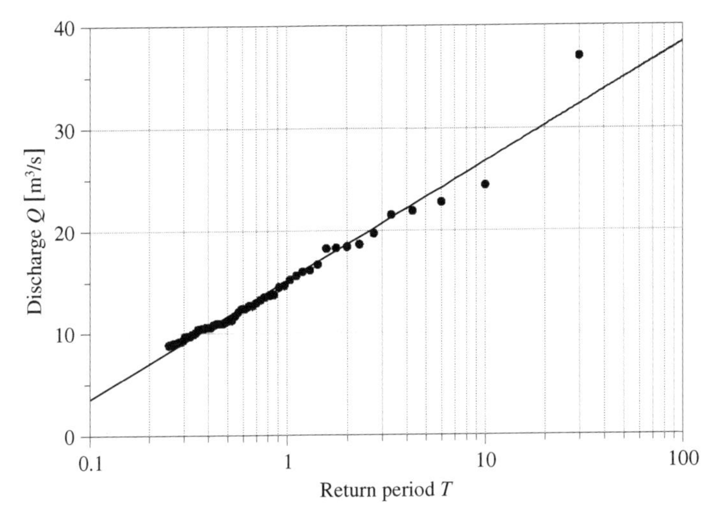

Fig. 8.6 Calibration of an exponential equation for a POT series for the Nant d'Avril River.

In this case we can fit the exponential equation (see below) to specify the two parameters a and b.

Example

Let us consider a POT series of 60 large floods over a 15-year period for the Nant d'Avril River.[25] Figure 8.6 shows the result of the fit using an exponential equation on a logarithm scale.

Apart from the simplicity of this method (in this case we have applied the regression technique), it should be noted that the return periods less than or equal to one year become available.

Modeling Using the Generalized Pareto Equation

This equation is written as:

$$x = a + \frac{b}{c}\left\{1 - \left(\frac{1}{T}\right)^{c}\right\}$$
$$T = \frac{1}{\left\{1 - \frac{(x-a)}{b}\,c\right\}^{\frac{1}{c}}} \tag{8.33}$$

To estimate the parameters we can use nonlinear regression software (see for example, Sherrod, 2002). Figure 8.7 shows the calibration of a POT series for the Nant d'Avril River.

[25] Data from the Canton of Geneva, Département de l'intérieur et de la mobilité, Direction générale de l'eau.

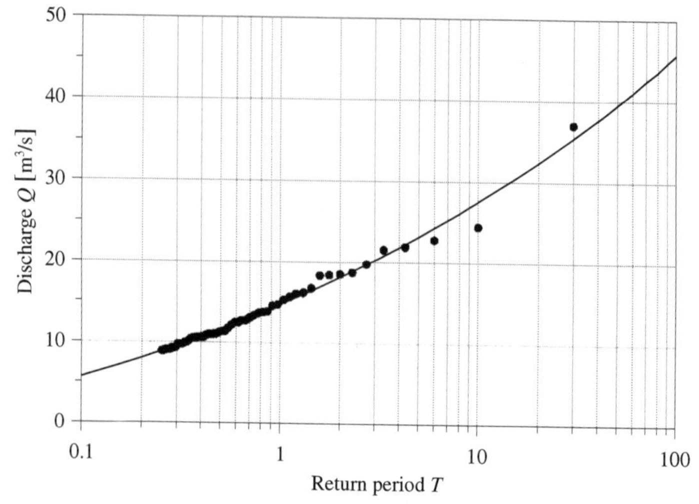

Fig. 8.7 Calibration of a POT series for the Nant d'Avril River using a generalized Pareto equation.

Confidence Interval

Since with these two equations we are unable to apply the classical techniques to determine the confidence interval (see Section 7.2), we will use an alternate method, the *parametric bootstrap* (See Section 7.4).

Meylan (2010) proved that it is sufficient to define the *standard error* of the value $x(T)$ under study between the limits $-2 \leq z \leq 2$ because the underlying sampling distribution is *normal* (see Fig. 8.8). For practical applications this technique is entirely satisfying. Based on this standard error it is then possible to compute again the confidence interval for any level of $1-\alpha$. Figure 8.9 shows the 80% confidence interval for the return periods of the Nant d'Avril River.

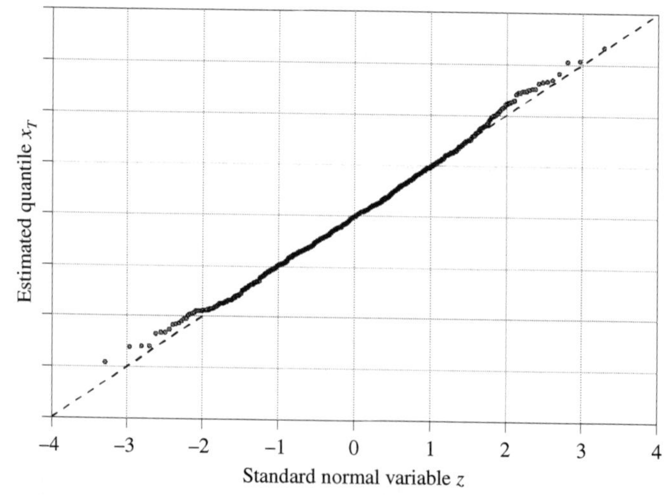

Fig. 8.8 Experimental sampling distribution for a quantile x_T.

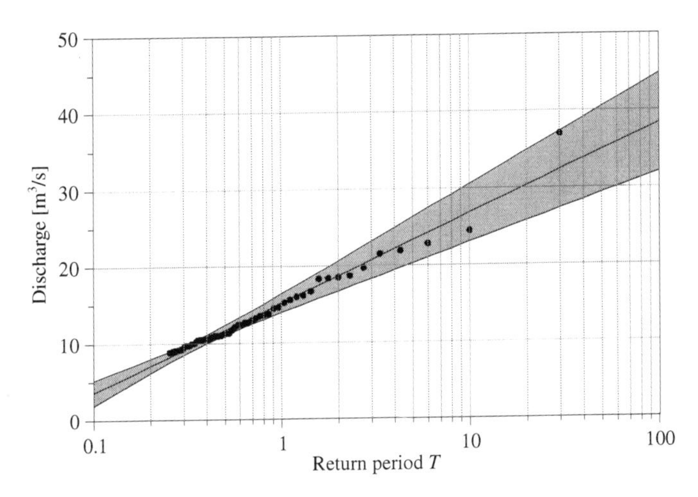

Fig. 8.9 Confidence interval at level $(1 - \alpha) = 80$ % for the Nant d'Avril River data (see Fig. 8.6).

8.6.4 Relationship to Classical Analysis

The Poissonian counting approach is asymptotically equivalent to the classical approach involving the annualization of a POT series (see Section 8.2.2 and Figure 8.10).

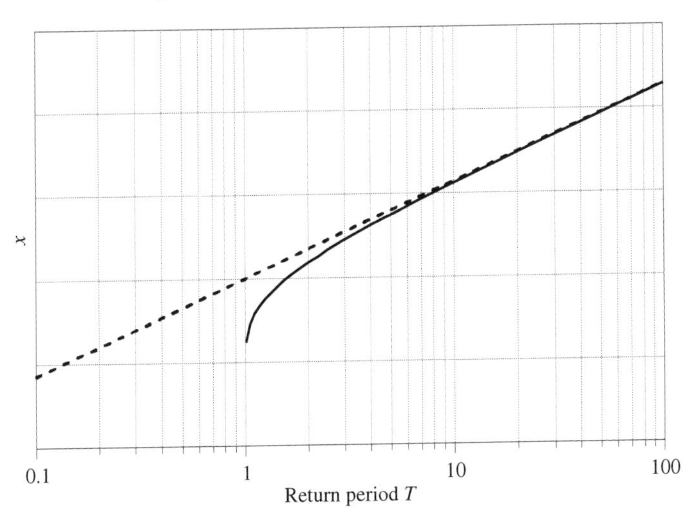

Fig. 8.10 Relation between the classical approach (Gumbel distribution in solid line) and the Poissonian counting approach (broken line).

In practice the equivalence is for $T > 10$ years. This equivalence makes it easy to move from a Gumbel distribution to the exponential equation as follows:

$$x(T) = \alpha_{\text{gum}} + \beta_{\text{gum}} \ln(T) \tag{8.34}$$

Similarly, we can move from a generalized extreme value distribution (GEV) to the generalized Pareto equation using:

$$x(T) = \alpha_{\text{gev}} + \frac{\beta_{\text{gev}}}{\gamma_{\text{gev}}}\left[1 - \left(\frac{1}{T}\right)^{\gamma_{\text{gev}}}\right] \tag{8.35}$$

The point process characterization of extremes is explained in detail in Coles (2001).

CHAPTER 9

Perspectives

This chapter deals with questions that are currently widespread in frequency analysis. More specifically, we will tackle multivariate frequency analysis using copulas and non-stationarity. We have also included a less recent field concerning regional frequency analysis (GREHYS, 1996) which is still under study. First, however, it would be worth taking a look at Bayesian statistics, although in frequency analysis, this approach remains somewhat marginal (Section 9.1).

9.1 BAYESIAN FREQUENCY ANALYSIS

9.1.1 The Principles of Bayesian Analysis

In the Bayesian approach, the values that the parameters $\theta = (\theta_1, \theta_2, \cdots, \theta_p)$ defining the model can take are uncertain rather than fixed, the opposite of the situation in classical statistics (also known as *frequentist statistics*). In Bayesian analysis, the uncertainty concerning the parameters is represented by a probability distribution called *prior* or *a priori* which is denoted $p(\theta)$. This distribution is established based on the information available *a priori* that does not result from a series of observations $x = (x_1, \ldots, x_n)$, but instead comes from other sources that can be either subjective (expert's or manager's knowledge...) or objective (previous statistical analyses). This means that the Bayesian approach has the advantage of formally incorporating into the analysis the knowledge that is available regarding the parameters of interest.

Once the available observations and the *a priori* distributions have been specified, Bayes' theorem is applied to combine the *a priori* information about the parameters with the information contained in the data, using the likelihood $p(x|\theta)$. Bayes' theorem is expressed as:

$$p(\theta|x) = \frac{p(x|\theta)p(\theta)}{\int p(x|\theta)p(\theta)d\theta}$$

Thus Bayes' theorem operates as an "information processor" to update the *a priori* knowledge in light of the observations (Bernier *et al.*, 2000).

The diagram of Figure 9.1 illustrates the basic steps in Bayesian reasoning that lead to an *a posteriori* inference.

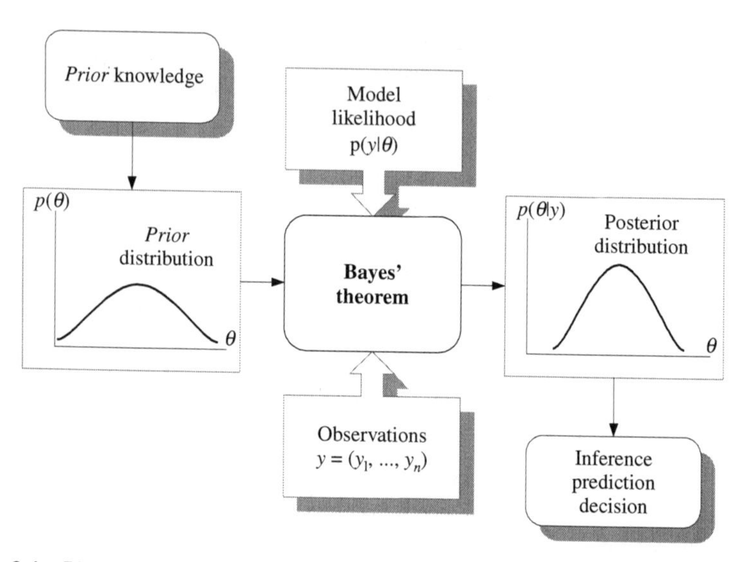

Fig. 9.1 Diagram illustrating the basic steps in Bayesian analysis (taken from Perreault, 2000).

The theoretical principles of Bayesian statistics are described in detail in the reference books by Berger (1985), Bernardo and Smith (1994) and Parent and Bernier (2007). For details on more practical aspects of applying the Bayesian approach, the reader is referred to Gelman *et al.* (2004) and Bernier *et al.* (2000).

9.1.2 The Advantages of the Bayesian Approach in Frequency Analysis

Several of the advantages of the Bayesian method in frequency analysis listed below are borrowed from Perreault (2000, 2003).

- Bayesian inference makes it possible to compute the joint probability (the *a posteriori distribution*) of all the unknown parameters, which in turn allows us to reach several conclusions simultaneously. We can not only obtain estimations of the parameters or functions of these quantities, such as percentiles (which are the values of interest in frequency analysis), but we can also evaluate their accuracy directly. It should be highlighted that unlike most classical methods of frequency analysis, Bayesian inference is not based on any asymptotic assumption. As Katz *et al.* (2002) pointed out, the standard error of parameters estimated using the maximum likelihood method can sometimes not be applied in practice because of the asymptotic assumptions on which this frequentist approach is based.

- Hydrologists often have unquantified information about the phenomenon of interest before developing a project. The Bayesian approach allows them to formally include their own expertise in their considerations by choosing *a priori distributions*. The possibility to combine this information with the data is even more important because in hydrology, the samples are usually of limited size.
- Managers who are required to make decisions based on the results of a frequency analysis find several advantages in applying the Bayesian method. The idea of updating the *a priori* knowledge of the experts based on information available in the data can be very interesting and reassuring to them. In addition, they are sensitive to the socio-economic aspects of managing water resources. As a consequence, they think that statistical analyses and the assessment of the socio-economic consequences of their decisions should not be handled separately (Bernier *et al.*, 2000). Now Bayesian analysis makes it possible to properly integrate the models in a whole decision-making process. This only requires that the analysis includes a function that quantifies the loss incurred for the different options. This is called a "complete" Bayesian analysis, the objective being to choose the optimal solution among all the possibilities.
- Historical data are often available for the estimation of events such as floods (see Section 3.2.5). This information is often of poorer quality than the data from more recent flow measurements. The Bayesian approach offers a framework that is well suited to incorporate these less reliable data.
- As we have already indicated, the result of a Bayesian analysis, corresponding to an *a posteriori* distribution, provides a more complete inference than the corresponding analysis based on maximum likelihood. More precisely, since the objective of an analysis of extreme values is usually to estimate the probability that future events reach a certain extreme level, it seems natural to express this quantity by means of a predictive distribution (Coles, 2001, Section 9.1.2).

9.1.3 Applications of Bayesian Inference in Frequency Analysis

The absence, in most practical cases, of explicit analytical solutions for *a posteriori* distributions, combined with the complexity of the resulting numeric computations, meant that the Bayesian method fell into disfavor. But recent advances in the field of numerical computation, which apply Markov chain Monte Carlo methods (MCMC) to implement Bayesian methods, have overcome this drawback (Robert and Casella, 2004). This has led to a widespread use of the Bayesian approach in various areas of statistical application (Berger, 1999). Surprisingly, the application of Bayesian methods within the water sciences community remains quite limited, contrary to other fields (medicine, finance, economics and biology, just to name a few). However, as we have already pointed out, the Bayesian perspective is conceptually closer to the concerns of the manager in water sciences than the so-called "classical" approach.

In hydrology, the analysis of flood probabilities (the estimation of percentiles) from a Bayesian perspective has not been the subject of much attention. Some of the first work in this regard includes Kuczera (1982, 1999) and Fortin *et al.* (1998). Coles and Tawn (1996) applied the Bayesian technique of elicitation of expert opinion[26] to integrate the knowledge of the hydrologist into the frequency analysis of maximum precipitations. This makes it possible to obtain a credibility interval (equivalent to the frequentist confidence interval) for the quantile corresponding to a 100-year return period. Parent and Bernier (2003a) developed a Bayesian estimation of a peak over a threshold model (POT) by incorporating an informative semi-conjugate *a priori* distribution. This approach makes it possible to avoid using MCMC algorithms, which can be difficult to implement. Katz *et al.* (2002) cite other applications of the Bayesian approach to frequency analysis.

9.2 MULTIVARIATE FREQUENCY ANALYSIS USING COPULAS

9.2.1 Using Copulas in Hydrology

In many areas of statistical applications, including hydrology, the analysis of multivariate events is of particular interest. For example, dam builders

[26] In Bayesian inference the term "elicitation" refers to the action of helping an expert to formalize his knowledge in order to define the *prior distribution*. The steps in an elicitation procedure include: definition of the quantities to elicit, the so-called elicitation itself, which is done through discussion and questions with the expert, the translation of this knowledge into the *a priori* distributions, and finally the verification by the expert of the concordance between these distributions and his own experience.

need to design structures as a function of river discharge, and this discharge increases considerably during the larger flows of springtime. Flows are usually described in terms of three main characteristics: peak flow, volume and duration. Because these three variables are correlated, three univariate frequency analyses cannot provide a full assessment of the probability of occurrence of the event in question. In addition, a univariate frequency analysis can overestimate the severity of a particular event (Yue and Rasmussen, 2002) and as a consequence, raise the construction costs of the structure.

Another application of multivariate frequency analysis in hydrology is to combine the risk downstream from the confluence of several rivers or from a cascade of several sub-watersheds. For several applications, the peak discharge is the result of the combination of discharges in several intermediate watersheds. Thus it is important to take into account any dependence between discharges.

The mathematical theory of univariate models for extreme events has been well established. Based on this theory, it is possible to develop multidimensional models for extreme values, which are the boundary distributions of the joint marginal distribution of the maxima.

Hydrologists apply several classical distributions for the multivariate analysis of extreme events, but such an approach is rarely efficient, partly because the available multivariate models are not very well suited to the representation of extreme values. For a long time, the normal model dominated statistical studies of multivariate distributions. This model is attractive because the conditional and marginal distributions are also normal. However this model is limited and thus it was necessary to find alternatives to the approach using a normal distribution. An abundant statistical literature exists dealing with multivariate distributions.

However, many multivariate distributions are direct extensions of univariate distributions (the bivariate Pareto distribution, the bivariate gamma distribution, etc.). These distributions have the following drawbacks:

- Each marginal distribution belongs to the same family.
- Extensions beyond the bivariate case are not clear.
- The marginal distribution parameters are also used to model the dependence between the random variables.

In hydrology, the most commonly applied multivariate distributions are the multivariate normal, bivariate exponential (Favre *et al.*, 2002), bivariate gamma (Yue *et al.*, 2001), and bivariate extreme value distributions (Adamson *et al.*, 1999). In the case of the multivariate normal distribution, the measure of dependence is summarized in the correlation matrix. In most cases, using a multivariate normal distribution is not appropriate for modelling maximum discharges as the marginal distributions are asymmetric and heavy-tailed. In addition, the dependence structure is usually different than the Gaussian structure described with Pearson's correlation coefficient. It should also be noted that in the case of more complex marginal distributions, such as mixtures of distributions that are widely used in the practice of modelling heterogeneous phenomena, it is not possible to apply standard multivariate distributions.

One construction of multivariate distributions that does not suffer from the drawbacks already mentioned is the concept of copulas. A copula is extremely useful for implementing efficient and realistic simulation algorithms for joint distributions. Copulas are able to model the dependence structure independently of the marginal distributions. It is then possible to build multidimensional distributions with different margins, the copula mathematically formalizing the dependence structure. The crucial step in the modeling process depends on the choice of the copula function that is best suited to the data and on the estimation of its parameters. Copulas have been widely used in the financial field as a way of determining the "Value at Risk" (VaR) (for example, Embrechts *et al.*, 2002, 2003; Bouyé *et al.*, 2000). Other areas of application include the analysis of survival data (Bagdonavicius *et al.*, 1999) and the actuarial sciences (Frees and Valdez, 1998). However in the field of hydrology, application of copulas still remains marginal.

Hydrologists are concerned with determining statistical quantities such as joint probabilities, conditional probabilities and joint return periods. Once a copula family has been selected, its parameters estimated and the goodness-of-fit tested, these quantities can be computed very simply and directly from the copula expression. Yue and Rasmussen (2002) and Salvadori and De Michele (2004) showed how to calculate these probabilities in the case of a bivariate distribution. Generalization in the multivariate case is straightforward.

The theory of copulas is available in general reference books such as Joe (1997) and Nelsen (2006). Genest and Favre (2007) introduced the required steps for multivariate modeling using copulas in a didactic way, with an emphasis on hydrological applications. Consequently, here we will provide only a quick summary of the theory of copulas.

9.2.2 Definition and Properties of Copulas

Definition

A copula in dimension p is a multivariate distribution C, with uniform marginal distributions on the interval [0,1] ($U[0,1]$) such that

- $C : [0,1]^p \rightarrow [0,1]$;
- C is bounded and p-increasing
- The marginal distributions C_i of C are such that: $C_i(u) = C(1,\ldots,1,u,1,\ldots,1) = u$ whatever $u \in [0, 1]$.

It can be deduced from the definition that if $F_1,\ldots,F_p$ are univariate distributions, then $C(F_1(x_1),\ldots,F_p(x_p))$ is a multivariate distribution with marginal distributions $F_1,\ldots,F_p$, since $F_i(x_i)$, $i = 1,\ldots,p$ is distributed according to a uniform distribution. Copulas are very useful for the construction and simulation of multivariate distributions. In simplified form, a copula can be viewed as a transfer function that makes it possible to link the marginal distributions and the multivariate joint distribution.

Figure 9.2 illustrates a copula in the form of a transfer function.

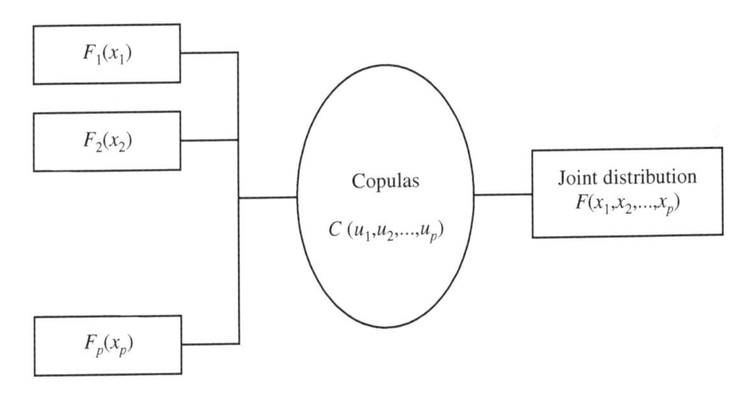

Fig. 9.2 Illustration of copulas as a transfer function.

The following theorem (Sklar, 1959) highlights the high potential of copulas for the construction of multivariate distributions.

Sklar's Theorem

Let F denote a probability distribution in dimension p with continuous marginal distributions $F_1,\ldots,F_p$; then F has the following unique representation in the form of a copula:

$$F(x_1,\ldots,x_p) = C\ (F_1(x_1),\ldots,F_p(x_p)) \tag{9.1}$$

It can be deduced from Sklar's theorem that it is possible to separate the marginal distributions and the multivariate dependence structure.

Schweizer and Wolff (1981) showed that copulas are able to take into account the whole dependence between two random variables X_1 and X_2 in the following way: let g_1 and g_2 be two strictly increasing functions on the domain of definition of X_1 and X_2. Thus the variables obtained by the transformation, namely $g(X_1)$ and $g(X_2)$ have the same copula as X_1 and X_2. This means, as stated by Frees and Valdez (1998), that the way in which X_1 and X_2 are linked is included in the copula, independently of the measurement scale of each variable. Schweizer and Wolff (1981) also proved that two classical nonparametric measures of correlation can be simply expressed by means of copulas – Kendall's tau and Spearman's rho.

9.2.3 Types of Copulas

Archimedean Copulas

There are several families of copulas applied for modeling a set of variables. The Archimedean class of copulas, popularized by Genest and MacKay (1986), is the family most widely applied in practice, due to its interesting analytical properties and the ease with which it can be simulated.

A copula C in dimensions p is said to be Archimedean if there exists a strictly decreasing continuous function φ: $[0, 1] \rightarrow [0, \infty]$ satisfying the condition $\varphi(1) = 0$ so that

$$C(u_1,\cdots,u_p) = \varphi^{[-1]}\{\varphi(u_1)+\ldots+\varphi(u_p)\}$$

where φ^{-1} represents the inverse of the generator φ.

The most common examples of Archimedean copulas are the Gumbel, Frank and Clayton families. The Clayton model is frequently applied in life cycle analysis, under the name gamma frailty model; see Oakes (1982) for example.

Table 9.1 summarizes the main Archimedean copulas and their generators in the bivariate case. The generator determines the Archimedean copula in a unique way (they can only differ by a constant).

Table 9.1 Main bivariate Archimedian copulas and their generators.

Family	Generator $\varphi(t)$	Parameter α	Bivariate copula $C_\varphi(u_1, u_2)$
1. **Independence**	$-\ln t$	–	$u_1 u_2$
2. **Clayton** (1978) Cook and Johnson (1981) Oakes (1982)	$t^{-\alpha} - 1$	$\alpha > 0$	$(u_1^{-\alpha} + u_2^{-\alpha} - 1)^{-1/\alpha}$
3. **Gumbel** (1960) Hougaard (1986)	$(-\ln t)^{\alpha}$	$\alpha \geq 1$	$\exp\left[-\{(-\ln u_1)^{\alpha} + (-\ln u_2)^{\alpha}\}^{1/\alpha}\right]$
4. **Frank** (1979)	$\ln\left\{\frac{\exp(\alpha t) - 1}{\exp(\alpha) - 1}\right\}$	$\alpha \neq 0$	$\frac{1}{\alpha}\ln\left[1 + \frac{\{\exp(\alpha u_1) - 1\}\{\exp(\alpha u_2) - 1\}}{\exp(\alpha) - 1}\right]$

The family of Archimedean copulas also includes the Galambos copula (1975), the Ali-Mikhail-Haq copula (1978), the Cuadras-Augé copula (1981), the Hüsler-Reiss copula (1989) and the Genest-Ghoudi copula (1994).

Meta-elliptical Copulas

The elliptical class of copulas holds considerable practical interest, because they can be easily applied in dimension $p\,(p>2)$. In addition, this class constitutes a generalization of the classical multivariate normal distribution. More precisely, a distribution is said to be elliptical if it can be represented in the form of

$$Z_{p\times 1} = \mu_{p\times 1} + RA_{p\times p}u_{p\times 1} \sim EC_p(\mu, \Sigma, g),$$

where μ is a location vector, R is a positive random variable, $\mathbf{A}$ is a $p \times p$ matrix such as $\mathbf{AA}^{\mathrm{T}} = \Sigma$, $\mathbf{u}$ is a uniformly distributed vector on the sphere of dimension p and g is a scale function.

In the particular case where $g(t) \propto e^{-t/2}$, we find the classical multivariate normal distribution. Other generators correspond to multivariate Student, Cauchy and Pearson type II distributions, among others (see e.g. Fang *et al.*, 2002 or Abdous *et al.*, 2005).

The marginals of elliptical distributions are entirely determined by the generator g and, moreover, are all identical. However, by transforming the variables individually to make the marginals uniform, we obtain a copula known as meta-elliptical, into which we can now insert the desired margins.

In other terms, it is possible to build a vector $X_{p\times 1} = (X_1,\cdots,X_p)$ with marginals $F_1,\ldots,F_p$ from an elliptical vector $(Z_1,\ldots,Z_p)$ by using

$$X_i = F_i^{-1}\{Q_g(Z_i)\},\ 1 \le i \le p$$

where Q_g is the common marginal of the variables $Z_1,\ldots,Z_p$ and F_i^{-1} denotes the inverse of the distribution function of the marginal F_i of X_i. This construction, based on a meta-elliptical copula, is written as

$$X_{p\times 1} \sim ME_p(\mu, \Sigma, g; F_1,\cdots,F_p).$$

Fang *et al.* (2002) and Abdous *et al.* (2005) discuss various properties of meta-elliptical copulas. Among these, the most important is the fact that Kendall's tau (τ) between two components of an elliptical vector is independent of the function g and is linked to the correlation r between the two variables by the formula $r = \sin(\tau\pi/2)$.

Genest *et al.* (2007) developed a trivariate frequency analysis of flood peak, volume and duration by applying meta-elliptical copulas.

9.2.4 Estimation of Copula Parameters

Without loss of generality, let us suppose that the dimension $p = 2$ and that a parametric family of copulas (C_θ) has been chosen to model the dependence between two random variables X and Y. Given a sample $(X_1,Y_1),\ldots,(X_n,Y_n)$ from (X,Y), how should the parameter θ be estimated?

Several methods exist for the estimation. The most direct approaches involve inverting Kendall's τ or Spearman's ρ. These two methods can only be applied if there is a direct relation between these nonparametric measures of dependence and the parameter of the copula. In addition, they can only be used if the parameter is a real number (and therefore unidimensional). Below we present an efficient approach that is based solely on ranks. This approach follows Genest's school of thought (discussed for example in Genest and Favre, 2007) which advocates ranks as the best summary of the joint dependence of random pairs.

Maximum Pseudolikelihood Method

In classical statistics, the maximum likelihood method is a well-known alternative to the method of moments and is usually more efficient, especially when θ is multidimensional. In the current case, an adaptation of this method is required, as we would like the inference concerning the dependence parameters to be based exclusively on ranks. Such an adaptation was described briefly by Oakes (1994) and was later formalized and studied by Genest *et al.* (1995) and by Shih and Louis (1995).

The maximum pseudolikelihood method assumes that C_θ is absolutely continuous with density c_θ and involves the maximization of the log-likelihood based on ranks in the following form:

$$l(\theta) = \sum_{i=1}^{n} \log\left\{c_\theta\left(\frac{R_i}{n+1}, \frac{S_i}{n+1}\right)\right\},$$

where R_i and S_i are the respective ranks of X_i and Y_i, $i = 1,\ldots,n$.

The above equation corresponds exactly to the expression obtained when the unknown marginal distributions F and G in the classical log-likelihood method

$$l(\theta) = \sum_{i=1}^{n} \log[c_\theta\{F(X_i),G(Y_i)\}]$$

are replaced by rescaled versions of their empirical counterparts, i.e.:

$$F_n(x) = \frac{1}{n+1}\sum_{i=1}^{n} 1(X_i \le x) \text{ and } G_n(y) = \frac{1}{n+1}\sum_{i=1}^{n} 1(Y_i \le y),$$

where 1 denotes the indicator function equal to one when the inequality holds and zero otherwise.

In addition, it is immediate that $F_n(X_i) = R_i/(n + 1)$ and $G_n(Y_i) = S_i/(n + 1)$ for all $i \in \{i,\ldots,n\}$.

9.2.5 Goodness-of-fit Tests

Recently, goodness-of-fit tests for copulas have been developed. But although the field is in constant development, only a few papers have actually been published in the literature on this topic. Goodness-of-fit tests for copulas fall into three general categories:

- goodness-of-fit tests based on the probability integral transformation, for example Breymann *et al.* (2003);
- goodness-of-fit tests based on the kernel estimation of the copula density, for example Fermanian (2005);
- goodness-of-fit tests based on the empirical process of copulas, for example Genest *et al.* (2006) and Genest and Rémillard (2007).

We prefer the third type of tests. The first implies conditioning on successive components of the random vector and has the drawback of depending on the order in which this conditioning is done. The second category of tests depends on various arbitrary choices, such as the kernel, the window size and the weight function, which make their application cumbersome.

9.2.6 Application of Copulas for Bivariate Modeling of Peak Flow and Volume[27]

A flow is composed of three main characteristics: the peak, the duration and the volume. Univariate frequency analyses of these quantities result in the over-estimation or under-estimation of risk (De Michele *et al.*, 2005), which can have disastrous consequences. The following example concerns the bivariate analysis of peak flows (maximum annual daily discharge) and the corresponding volumes for the Harricana River. The watershed, with a area of 3,680 km^2, is located in the northwest region of the province of Quebec, Canada. The data considered for this application include the annual maximum discharge X (in m^3/s) and the corresponding volume Y (in hm^3) for 85 consecutive years starting in 1915 and ending in 1999. The peak flow follows a Gumbel distribution with mean 189 [m^3/s] and standard error 51.5 [m^3/s]. The volume is adequately modeled by a gamma distribution with mean 1043.88 [hm^3] and standard error 234.93 [hm^3].

[27] A major portion of this section is taken from Genest and Favre (2007).

The scatter plot of ranks shown in Figure 9.3 suggests a positive association between peak flow and volume as measured by Spearman's rho (which represents the correlation of ranks). Computation of the two standard nonparametric measures of dependence gives us $\rho_n = 0.696$ for Spearman's rho and $\tau_n = 0.522$ for Kendall's tau.

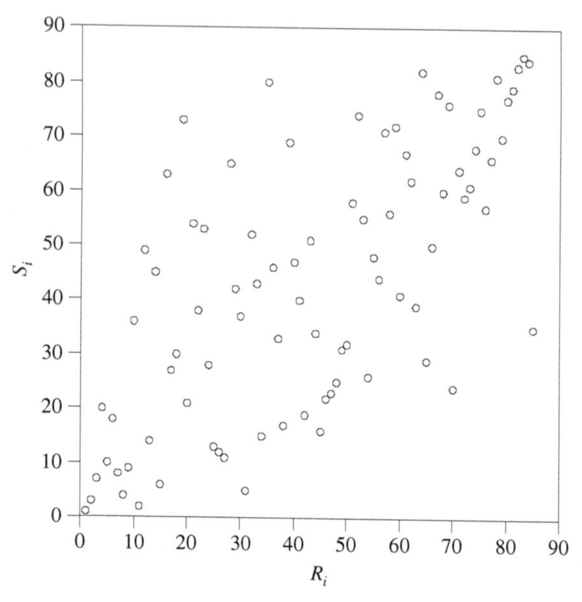

Fig 9.3 Scatter plot of the ranks of peak flow (R_i) and corresponding volume (S_i).

In order to model the dependence between the annual peak and the volume, some 20 families of copulas were considered. These could be classified into four broad categories:

1. Archimedean copulas with one, two or three parameters, including the traditional Ali-Mikhail-Haq (Ali *et al.*, 1978), Clayton (1978), Frank (Nelsen, 1986; Genest, 1987) and Gumbel-Hougaard (Gumbel, 1960) families. The families of Kimeldorf and Sampson (1975), the class of Joe (1993) and the BB1-BB3, BB6-BB7 described in the book by Joe (1997, p. 150-153) were also considered as potential models.
2. Extreme values copulas, including (besides the Gumbel-Hougaard copula mentioned above) Joe's BB5 family and the classes of copulas introduced by Galambos (1975), Hüsler and Reiss (1989) and Tawn (1988).
3. Meta-elliptical copulas, more specifically the normal, Student and Cauchy copulas.

4. Various other families of copulas, such as those of Farlie-Gumbel-Morgenstern and Plackett (1965).

The Ali-Mikhail-Haq and Farlie-Gumbel-Morgenstern families could be eliminated off hand because the degree of dependence they span were insufficient to account for the association that we can observe in the data set. To discriminate between the other models we used a graphical tool described in detail Genest and Favre (2007). This graphic is constructed as follows: given a family of copulas (C_θ), an estimation θ_n of its parameters is obtained using the maximum pseudolikelihood method. Then 10,000 pairs of points (U_i, V_i) were generated from C_{θ_n}. The resulting pairs were transformed back into their original units by applying the inverse of the marginal distributions identified beforehand (as mentioned earlier, the domain of definition of a bivariate copula is the unit square, $[0,1]\times[0,1]$). Figure 9.4 shows a scatterplot of these pairs with the observations superimposed for the five best families as well as the traditional bivariate normal model.

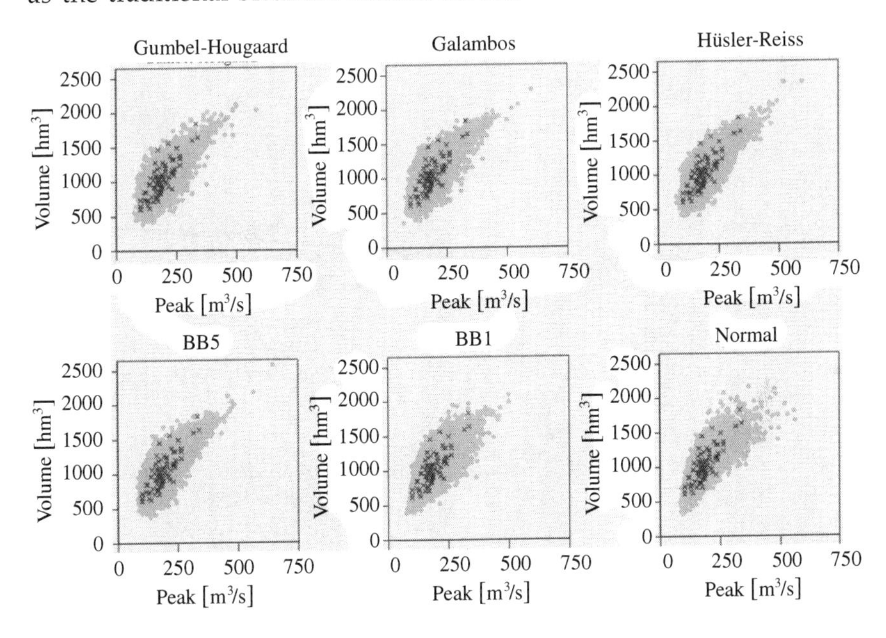

Fig. 9.4 Simulations of 10,000 peak flows and corresponding volumes using various families of copulas (Gumbel-Hougaard, Galambos, Hüsler-Reiss, BB5, BB1 and normal) and marginal distributions (Gumbel for the peak and gamma for the volume). The observations are superimposed, and indicated with the symbol *x*.

Keeping in mind the predictive ability that the selected model should possess, the normal copula was rejected due to the obvious lack of fit in the upper part of the distribution. Table 9.2 shows the definition of the five chosen copula models, as well as the domain of definition of their parameters. Note that four of these families of copulas belong to the extreme values class (Gumbel-Hougaard, Galambos, Hüsler-Reiss and BB5). The fifth (BB1) belongs to the two-parameter Archimedean class of copulas.

Table 9.2 Mathematical expressions of the five chosen families of copulas, with their parameter space

Copula	$C_\theta(u, v)$	Parameter(s)
Gumbel-Hougaard	$\exp\{-(\tilde{u}^{\theta} + \tilde{v}^{\theta})^{1/\theta}\}$	$\theta \geq 1$
Galambos	$uv\exp\{(\tilde{u}^{-\theta} + \tilde{v}^{-\theta})^{-1/\theta}\}$	$\theta \geq 0$
Hüsler-Reiss	$\exp\left[-\tilde{u}\Phi\left\{\frac{1}{\theta} + \frac{\theta}{2}\log\left(\frac{\tilde{u}}{\tilde{v}}\right)\right\} - \tilde{v}\Phi\left\{\frac{1}{\theta} + \frac{\theta}{2}\log\left(\frac{\tilde{v}}{\tilde{u}}\right)\right\}\right]$	$\theta \geq 0$
BB1	$[1 + \{(u^{-\theta_1} - 1)^{\theta_2} + (v^{-\theta_1} - 1)^{\theta_2}\}^{1/\theta_2}]^{-1/\theta_1}$	$\theta_1 > 0,\ \theta_2 \geq 1$
BB5	$\exp\left[-\{\tilde{u}^{\theta_1} + \tilde{v}^{\theta_1} - (\tilde{u}^{-\theta_1\theta_2} + \tilde{v}^{-\theta_1\theta_2})^{-1/\theta_2}\}^{1/\theta_1}\right]$	$\theta_1 \geq 1,\ \theta_2 > 0$

Note: with $\tilde{u} = -\log(u)$, $\tilde{v} = -\log(v)$ and Φ standing for the cumulative distribution function of the standard normal.

Table 9.3 contains the parameters estimation obtained with the maximum pseudo-likelihood method as well as the corresponding 95% confidence interval (Genest *et al.*, 1995) for each of the five chosen models.

Table 9.3 Parameters estimated using the-maximum pseudolikelihood method and the corresponding 95% confidence interval.

Copula	Estimated parameter(s)	95% confidence interval
Gumbel-Hougaard	$\hat{\theta}_n = 2.161$	IC = [1.867, 2.455]
Galambos	$\hat{\theta}_n = 1.464$	IC = [1.162, 1.766]
Hüsler-Reiss	$\hat{\theta}_n = 2.027$	IC = [1.778, 2.275]
BB1	$\hat{\theta}_{1n} = 0.418$, $\hat{\theta}_{2n} = 1.835$	IC = [0.022, 0.815] × [1.419, 2.251]
BB5	$\hat{\theta}_{1n} = 1.034$, $\hat{\theta}_{2n} = 1.244$	IC = [1.000, 1.498] × [0.774, 1.294]

Goodness-of-fit tests based on the empirical process of copulas were then applied in order to differentiate between the five families of copulas (see Genest and Favre (2007), p. 363 and Tables 12 and 13). These tests rely on the parametric bootstrap method and show that at the 5% level, none of the models still under consideration can be rejected. If the objective is to choose only one of the families, the one with the largest p-value in the goodness-of-fit test could be chosen. If we consider the test using the Cramér-von Mises process, the BB5 copula has the largest p-value ($p = 0.6362$, see Table 13 in Genest and Favre (2007)).

As an example, let us consider the following return period computation

$$T'(x, y) = \frac{1}{\Pr(X > x, Y > y)}$$

for various values of x and y corresponding to fixed univariate return periods. Now we have

$$\begin{aligned} \Pr(X_1 > x_1, X_2 > x_2) &= 1 - F_1(x_1) - F_2(x_2) + F(x_1, x_2) \\ &= 1 - F_1(x_1) - F_2(x_2) + C_\theta(u, v) \qquad (9.2) \\ \text{with } u &= F_1(x_1) \text{ and } v = F_2(x_2) \end{aligned}$$

Figure 9.5 illustrates the computation of this bivariate probability. Basically, we have: the area in (4) = 1 – area in (1) – area in (2) + area in (3) (equality translated as a probability in the first line of equation 9.2).

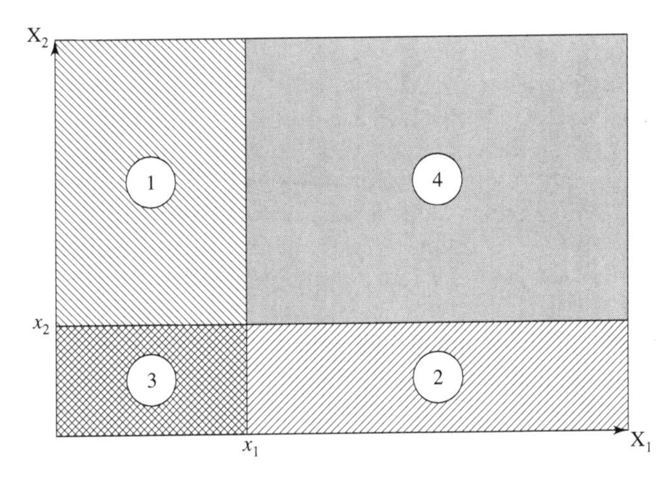

Fig. 9.5 Illustration of the computation of the bivariate probability $\Pr(X_1 > x_1, X_2 > x_2)$.

We denote $x_{[T]}$ (respectively $y_{[T]}$), the quantile corresponding to a peak flow (and respectively to a volume) of return period T. We applied equation 9.2 with the BB5 family of copulas and obtained the following bivariate return periods:

$T'(x_{[10]}, y_{[10]}) = 16.16$ [years], $T'(x_{[50]}, y_{[50]}) = 84.00$ [years] and
$T'(x_{[100]}, y_{[100]}) = 168.83$ [years]

If the peak flow and volume had been considered as independent, the return periods would have been $10^2 = 100$ [years], $50^2 = 2{,}500$ [years] and $100^2 = 10{,}000$ [years] respectively. By assuming such independence, the risk would have been highly underestimated.

This example clearly highlights the importance of taking into account the dependence between these two quantities, and in a more general sense, the importance of applying a multivariate approach in cases such as the one illustrated.

Favre *et al.* (2004) developed another application for copulas for the bivariate frequency analysis of peak flow and volume. In this context, Poulin *et al.* (2007) highlighted the importance of taking into account the dependence between the largest values (modeled in copula theory by the concept of tail dependence), mainly if extrapolation is required.

9.3 FREQUENCY ANALYSIS IN A NON-STATIONARITY CONTEXT

Classical frequency analysis is based on the assumption of underlying independent and identically distributed (i.i.d.) random variables. The second of these assumptions is not valid in a non-stationarity context, induced by climate changes for example. Basically, strict-sense stationarity means that the distribution remains constant over time. From a practical point of view, statisticians are used to relying on second-order stationarity, which implies that the first two moments (mean and variance) do not vary over time. The frequency analysis of a non-stationary series calls for a different understanding than the conventional approach involving stationarity. In fact, in the context of climate change, the distribution parameters and the distribution itself are likely to be modified. As a consequence, the exceedance probability used to estimate the return period also varies over time.

Several recent methods make it possible to take into account, at least partly, non-stationarity in the context of a frequency analysis.

The simplest method, which is not very recent (Cave and Pearson, 1914) involves first removing the trend, for example, by differentiating the series one or several times and then choosing the distribution and estimating its parameters. Below we give an overview of some recent approaches, taken mostly from a recent literature review (Khaliq *et al.*, 2006).

9.3.1 Method of Moments that Change Over Time

Strupczewski and Kaczmarek (2001) as well as Strupczewski *et al.* (2001 a, b) incorporate a linear or parabolic trend in the first two moments of the distribution. To apply this method, the distribution parameters are expressed as a function of the mean and the variance. The time-dependent parameters are estimated using the maximum likelihood method or the weighted least-squares approach. The validity of this method was proven by using non-negative asymmetric distributions with the two first moments expressed analytically as a function of the parameters. Six distributions were under consideration: the normal distribution, the 2-parameter lognormal, 3-parameter lognormal, gamma, Pearson type III and Gumbel distributions. For example, when considering a Gumbel distribution and a linear trend for both mean and variance, which is to say $\mu(t) = \mu_0 + \mu_1 t$ and $\sigma(t) = \sigma_0 + \sigma_1 t$, the Gumbel distribution parameters can be expressed as

$$\beta = \sqrt{6}\,\frac{\sigma_0 + \sigma_1 t}{\pi}, \quad \alpha = \mu_0 + \mu_1 t - 0.5772\beta$$

This means that the baseline distribution parameters are replaced with the following trend parameters (μ_0, μ_1, σ_0, σ_1). It is obvious that this approach generates a family of distributions, meaning that each time (t) has a distribution of its own.

9.3.2 Non-Stationary Pooled Frequency Analysis

Cunderlik and Burn (2003) propose a second-order non-stationary approach for flood frequency analysis by assuming non-stationarity in the first two moments, or in other words of the mean and variance of a time series. The quantile function at the local scale is splitted into a local non-stationary part that includes location and scale parameters and a regional stationary part. With this method the quantile function is expressed as $Q(F,t) = \omega(t)q(F)$ where $Q(F,t)$ is the quantile with probability F at time t, $\omega(t) = \mu(t)\sigma(t)$ is a local time-dependent component and $q(F)$ is a regional component independent of time. The parameters of the part dependent on time are estimated by splitting the time series into a trend portion and a time-dependent random variable representing the residual. This last variable models irregular fluctuations around the trend. The validity of this method is proven by assuming a linear trend for $\mu(t)$ and $\sigma(t)$.

Covariables Method

The statistical modeling of extremes is usually handled by one of the three main approaches: generalized extreme values distribution (GEV), peak over a threshold (POT, for example Reiss and Thomas, 2001) or point processes (for example Coles, 2001).

The idea underlying the covariables approach is to integrate changes that have taken place in the past directly into the techniques of frequency analysis in order to extrapolate in the future. It is possible that the statistical characteristics of extreme values vary as a function of low frequency climate indices such as El Niño-Southern Oscillation (ENSO). Modifications in the extreme values and their links with ENSO have been modeled by incorporating covariables into the distribution parameters (McNeil and Saladin, 2000; Coles, 2001; Katz *et al.*, 2002; Sankarasubramanian and Lall, 2003). Given a vector of covariables $V=v$, the conditional distribution of the maximum is assumed to be GEV with parameters dependent on v. For example, by considering the time as an explanatory variable, we can assume that the location parameter, the logarithm of the scale parameter and the shape parameter of the GEV distribution are linear functions of time. However, the shape parameter is often considered to be independent of time because the other two parameters are usually more important (McNeil and Saladin, 2000; Zhang *et al.*, 2004) and this parameter is also difficult to estimate even in the classical case of stationarity. Covariables can be inserted in the same manner for peak over a threshold models (McNeil and Saladin, 2000, Katz *et al.*, 2002) and in approaches based on point processes.

Approaches Based on Local Likelihood

Trend analysis is usually applied to determine an appropriate trend relation for frequency analysis in a context of non-stationarity. If *a priori* information about the structure of change exists, this approach can be better than defining a pre-specified parametric relation. In this context, the semi-parametric approach based on local likelihood (Davison and Ramesh, 2000; Ramesh and Davison, 2002) can be very useful for an exploratory analysis of extreme values. In these semi-parametric approaches, the classical models for reproducing the trends in annual maximum or POT series are estimated by applying a local calibration, while the parameters of these models are estimated separately for each time by weighting the data in the appropriate manner. This method leads to an estimation of parameters that is time-dependent and produces local estimations of extreme quantiles.

Estimation of Conditional Quantiles

In this type of approach, the conditional distribution of quantiles is studied in the presence of covariables or predictors, which means that the distribution of the parameters or of the moments is conditioned by the state of the covariables.

The best known method is based on quantile regression. This parametric approach was developed by Koenker and Basset (1978). It relies on the estimation of conditional quantiles by minimizing the weighted sum of asymmetric deviations by attributing different weights to the positive and negative residuals applying optimization techniques. Let p be the conditional quantile defined by the following regression: $Y_p(t) = \psi_p(V_t) + \varepsilon_p(t)$ where $\psi_p(.)$ is a linear or nonlinear function linking the p conditional quantile and the climate indices and $\varepsilon_p(t)$ is a white noise with zero mean and variance σ_p^2. The noise process can be either homoscedastic (with constant variance), or heteroscedastic (the variance depends on the explanatory variable). Let $V = (V_{1t}, V_{2t}, \ldots, V_{mt})$ be the vector of m covariables, corresponding for example to climate indices. The function ψ is obtained by resolving the following minimization problem:

$$\min_{\psi(V_t)} \sum_{t=1}^{n} R_p\left[Y_p(t) - \psi_p(V_t)\right] \text{ where } R_p(u) = \lfloor |\mu| + (2p - 1)u \rfloor / 2$$

If the regression function is assumed to be linear then for $p = 0.5$, the regression is defined by means of $\psi_p(V_t) = a_p(V_t)$ where a_p is a $m \times 1$ regression coefficient vector for the p-quantile. Koenker and d'Orey (1987) developed an algorithm for estimating a_p by using linear programming.

Another approach based on local likelihood makes it possible to insert covariables into the local estimation with the aim of estimating the conditional quantiles. For local likelihood, the weights are chosen by applying, for example, a kernel method (Sankarsubramanium and Lall, 2003).

Despite the existence of all these methods, a fundamental problem remains to be solved. The return period is defined as the mean of the recurrence interval, measured on a very large number of occurrences. Now, in a context of non-stationarity, the mean changes over time and as a consequence, the concept of return period makes no sense. As a result, serious thought is required to find a concept that can replace return period in a non-stationarity context.

9.4 REGIONAL FREQUENCY ANALYSIS

A regional analysis of hydrometeorogical variables can answer to two main needs:

- We need to be able to use these parameters on a spatial basis. We are thinking here especially of hydrometeorological data concerning rain or temperature, but also of relation parameters for the regionalization of discharges.
- We need robust local estimations, a characteristic that ideally can be improved by simultaneously taking into account the data from several stations in a single region.

We can catch a glimpse of the possibility of solving problems in areas where measurements are insufficient or nonexistent, and also the possibility of improving local estimations. The interactions between these approaches are straightforward and are illustrated in Fig. 9.6.

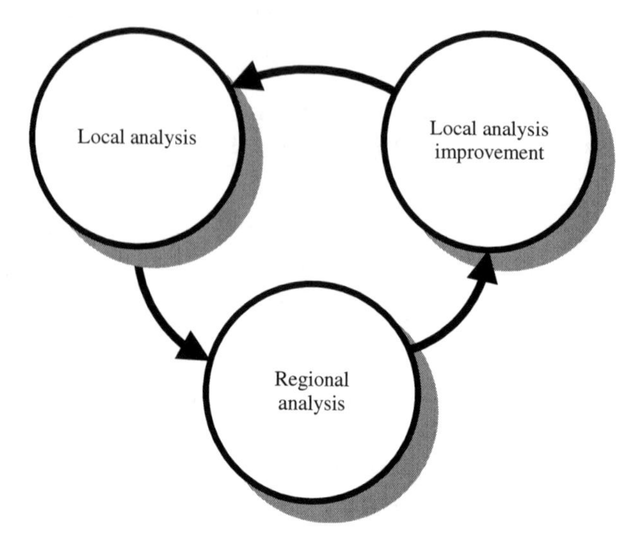

Fig. 9.6 Concept of regional analysis.

It is impossible to solve this problem in this section because the literature – and attempts at finding a solution – is too extensive. Also, research on this topic is still in fast development (Ouarda *et al.*, 2006; Castellarin *et al.*, 2007; Neykov *et al.*, 2007; Ribatet *et al.*, 2007; Viglione *et al.*, 2007). All we can accomplish here is to introduce a concept that is indispensable for hydrologists. Moreover it is important to be aware that in these techniques of regionalization, frequency analysis in the strictest sense plays a relatively modest role in comparison to all the other aspects of hydrology – rainfall-runoff modeling, for example.

We have chosen to divide the techniques of regional analysis into two main categories (GREHYS, 1996): the approach using the *anomaly method* and the one using *regional frequency analysis*. In both cases the basic hypothesis to be considered is that the phenomenon under study shows a certain degree of *homogeneity* over the region of interest. This hypothesis can be assessed either at the level of a coherent spatial structure of the variable of interest (residual method), or of the homogeneity of behavior allowing the introduction of auxiliary explanatory variables (for example, a definition of the flood index by a series of parameters), or even a combination of these two explanatory approaches.

9.4.1 Anomaly Method

The anomaly method (Peck and Brown, 1962) is based on the idea that the variable of interest is correlated to variables that are easy to define for each point in the region to be mapped (altitude, distance from a coast or a mountain chain, slope orientation, etc.).

We make the assumption that we have a sample of the variables of interest for N stations. Three phases are then involved:

1. Construction of an explanatory model

A multiple regression model for the response variable z can be constructed as follows:

$$\hat{z} = a_0 + a_1 w_1 + \cdots + a_p w_p \tag{9.3}$$

where $w_1,\ldots,w_p$ denote the p explanatory variables and $a_0,\ldots,a_p$ are the p+1 parameters estimated by least-squares using data from the stations $i = 1,\ldots,N$. For each station i the model implementation results in an anomaly (or residual):

$$r_i = \hat{z}_i - z_i \tag{9.4}$$

which models the difference between the value z_i observed at the station i and the value $\hat{z}_i$ estimated by the model.

After this first step we obtain a series of N residuals r_i (see for example Fig. 9.7) for which the variance is much lower than the variance of the basics field z_i.

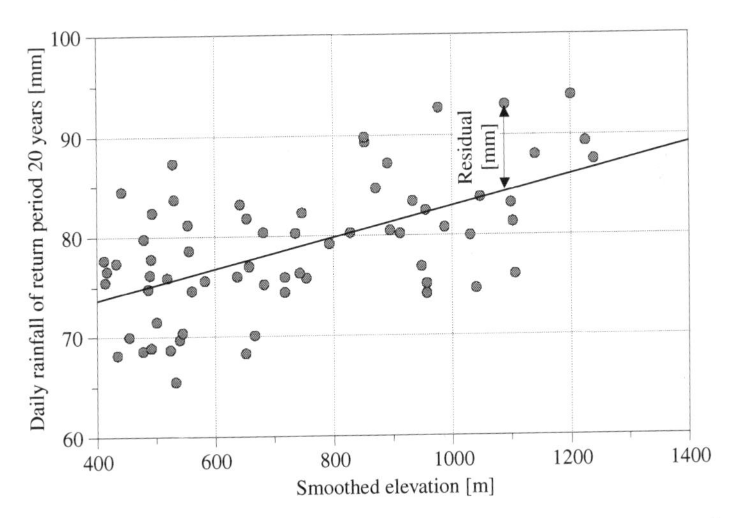

Fig. 9.7 Relation between daily rainfall corresponding to a 20-year return period and smoothed altitude for a window of 6 × 6 kilometers, based on Meylan (1986).

The several variants of the method differ mainly with regards to the choice of the explanatory variables.

2. Spatial interpolation of residuals

In this step, we must make the assumption that the residual field shows a coherent *spatial structure,* an assumption that can be verified by studying the *variogram* of residuals. It is then possible to regionalize the residuals on a regular grid, by applying the *kriging* method, for example.

3. Reconstruction of the variable of interest

For each point of the grid where it is possible to determine the p explanatory variables of the regression model, we can easily reconstruct the estimation of the variable of interest z as follows:

$$\hat{z}(x,y) = a_0 + a_1 w_1(x,y) + \cdots + a_p w_p(x,y) + r(x,y) \tag{9.5}$$

where $r(x, y)$ represents the value of the "spatial" residual at point (x,y).

Applications

The various applications of the anomaly method, the majority of which are applied for rainfall studies, can basically be distinguished by the choice of the explanatory variables.

De Montmollin *et al.* (1979) used smoothed altimetric information for a 500 meters square to model annual rainfall modules for the Mentue watershed.

Jordan and Meylan (1986 a, b) used smoothed altitude on a 6 km × 6 km window to study maximum daily rainfall in western Switzerland. In addition, a technique to take into account uncertainties in the data was applied by using a *structural variogram* (Meylan, 1986). Figure 9.7 shows the relation between daily rainfall corresponding to a 20-year return period and smoothed altitude, while Fig. 9.8 illustrates the result of regionalization.

Bénichou and Le Breton (1987) applied a principal components analysis of the landscape to select the explanatory variables for statistical rainfall fields. This approach made it possible to define the types of landscape that are most relevant for understanding rainfall fields: entrenchment effect, general west-east slope, general north-south slope, north-south pass, etc.

9.4.2 Regional Frequency Analysis

In the absence of a more highly developed approach, which does not yet exist, the *flood index* method is commonly used in practice to estimate flood discharge.

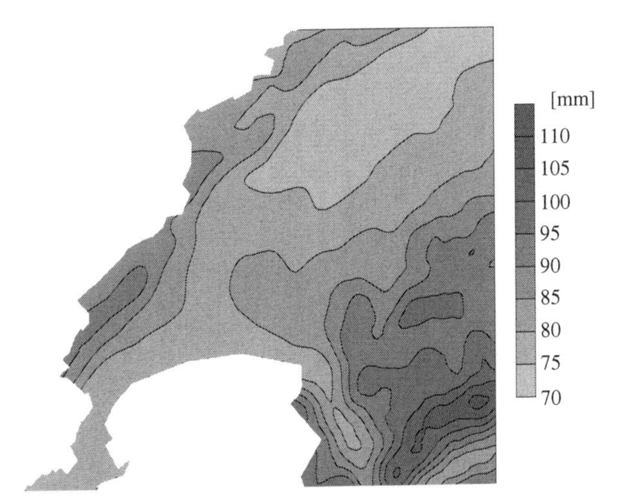

Fig. 9.8 Map of daily rainfalls [mm] corresponding to a 20-year return period for western Switzerland, based on Meylan (1986).

The flood index method was created for the purpose of regionalizing discharges (Dalrymple, 1960). It is usually used for large watersheds.

The basic hypothesis is that the region is homogenous, which is to say that the frequency distributions $Q_i(F)$ of maximum annual floods at each of the N stations i are identical, with only one different scale factor, which is called the *flood index*. Thus we can write:

$$Q_i(F) = \mu_i q(F),\ i = 1, \cdots, N \tag{9.6}$$

where μ_i, the mean maximum annual flood, is the flood index representing a scale factor,

$q(F)$ is the standardized regional frequency distribution.

It should be noted that $q(F)$ is also called the *region curve* (NERC, 1975) or *regional growth curve* (NERC, 1983) or also the *growing factor*. The general procedure introduced by Dalrymple (1960) remains one of the most popular for a regional approach to runoff (GREHYS, 1996).

The regional model has at least two parameters:

- The *flood index* μ_i, whose value varies from one measurement station to another and which must be regionalized if the goal is to apply the model for points where measurements are not available;
- the *growth curve* $q(F)$, identical for each point in each of the *homogeneous regions*.

The Hosking and Wallis Procedure

Hosking and Wallis (1993) proposed the following method for regional frequency analysis.

The mean of the flood discharges at station i expressed in equation (9.6) is given by $\hat{\mu}_i = \overline{Q}_i$, which is to say, the mean of Q_{ij}: annual flood discharges for the year j at station i. Other estimators can be considered, such as the median or the trimmed mean.

The following normed data

$$q_{ij} = \frac{Q_{ij}}{\hat{\mu}_i},\ j = 1, \cdots, n_i;\ i = 1, \cdots, N \qquad (9.7)$$

therefore serve as the basis for the estimation of $q(F)$. In equation (9.7), n_i denotes the total number of years of observations.

This method is based on the hypothesis that $q(F)$ is known and that only the p parameters $\theta_1,\ldots,\theta_p$ need to be estimated.

The procedure consists of estimating, separately for each station i, the parameters $\hat{\theta}_k^{(i)}$ which are the "local" parameters. Next, these parameters are combined as follows to give the regional estimation:

$$\hat{\theta}_k^{(R)} = \frac{\sum_{i=1}^{N} n_i \hat{\theta}_k^{(i)}}{\sum_{i=1}^{N} n_i} \qquad (9.8)$$

Equation (9.8) is a weighted mean, where the estimation for station i has a weight proportional to n_i (the sample size). These estimations lead to $\hat{q}(F)$, the regional growth curve.

Hosking and Wallis (1993) also proposed several scores that allow to check the data (using a measure of the *discordance* D_i), identify the homogeneous regions (with a measure of *heterogeneity* H) and choose the regional distribution to use (by way of a measure of *fit* Z).

Example of Application

Niggli (2004) studied the discharges of 123 rivers in western Switzerland and after dividing the area into three "homogeneous" regions and removing watersheds with an area less than 10 km^2, obtained the growth curves shown in Fig. 9.9.

To regionalize the flood indices, Niggli (2004) proposed the following equation for two of the regions, the *Plateau* and the *Pre-Alps*:

$$\mu = 0.0053A^{0.78}ALT^{0.68}\exp(0.58DRAIN - 0.025FOREST + 0.023IMP) \qquad (9.9)$$

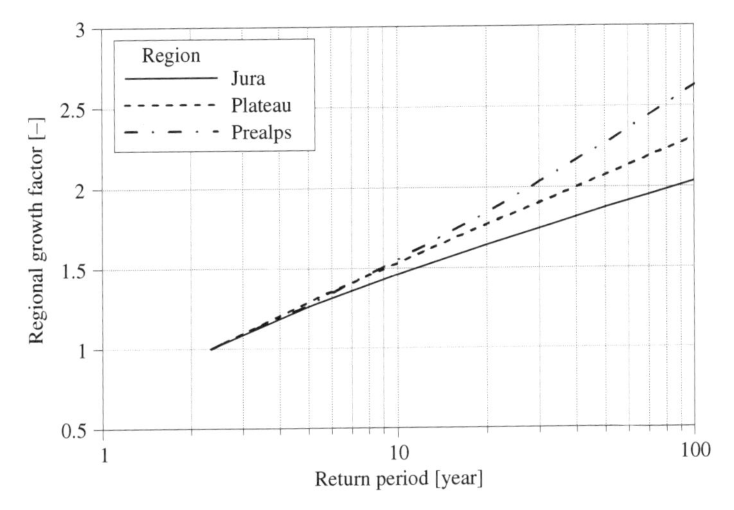

Fig. 9.9 Growth curve $q(T)$ for three "homogeneous" regions, namely the Jura, the Plateau and the Pre-Alps (Switzerland). Based on Niggli (2004).

where μ is the flood index under study,
A is the area of the watershed [km^2],
ALT is the mean altitude of the watershed [m],
DRAIN is the drainage density [km^{-1}] defined by L/A where L is the length of the hydrographic network,
FOREST is the proportion of forested area [%],
IMP is the proportion of impermeable area [%].

Thus Niggli (2004) applied a multiple linear regression with five explanatory variables selected on the basis of a "stepwise" regression procedure.

9.4.3 Other Approaches

This paragraph gives as an example an approach that is difficult to attribute to either the *anomaly method* or to *flood index* techniques.

Often, a technique for estimating flood discharge in small watersheds (generally those with an area less than 10 km^2 (OFEG, 2003)) is advocated. This kind of technique falls in between the anomaly methods and the flood indices methods. Anomaly methods focus essentially on the regionalization of residuals. The flood index methods look at a homogeneous region to construct a "global" frequency model. The intermediate method discussed here is based on a conceptual hydrological model, with some of the parameters being *explanatory variables* (slope, runoff coefficient, IDF) and others being calibration parameters validated for the entire region or that can be regionalized (see the parameters α and β below).

This means that the *Rational Formula,* for example, is written as:

$$Q(T) = uC_r i(T,d)A \tag{9.10}$$

where $Q(T)$ is the flood discharge corresponding to return period T [m^3/s],
C_r is the runoff coefficient of the watershed [-],
$i(T,d)$ is the rainfall intensity with return period T and duration d [mm/h],
d is the duration of the rain event involved, usually considered as equal to the concentration time t_c of the watershed [min],
A is the area of the watershed [km^2],

and $u = 0.278$ is a constant used to change the measurement units.

The intensity of rainfall with return period T and duration d can be described using a Montana formula (see equation (8.14)) rewritten as:

$$i(T, d) = \varepsilon_1(T)d^{\varepsilon_2(T)} \tag{9.11}$$

where $\varepsilon_1(T)$ and $\varepsilon_2(T)$ are the parameters of the Montana relation for the given return period T.

The most delicate aspect of this technique consists of formulating a relation for the concentration time t_c (which determines the duration d of the rainfall under study). Niggli (2004) proposed the following equation (9.12), which summarizes the basics of the different forms of this equation found in the literature:

$$t_c = \beta\left(\frac{A}{p}\right)^{\alpha} \tag{9.12}$$

where p is the mean slope of the watershed [%], and
α and β are two parameters to be estimated.

Based on equations (9.11) and (9.12), Niggli (2004) expresses the equation for the Rational Method (equation(9.10)) as follows:

$$Q(T) = uC_r\beta^{\varepsilon_2(T)}p^{-\alpha\varepsilon_2(T)}\varepsilon_1(T)A^{1+\alpha\varepsilon_2(T)} \tag{9.13}$$

This model has two calibration parameters, α and β. Niggli (2004) and several other authors note that the parameter α can often be considered as a constant and propose setting it equal to 0.5. Parameter β serves to characterize the watershed and can be regionalized or used to define *homogeneous regions.* Therefore and applying this principle, Niggli (2004), for example, has divided western Switzerland into three homogeneous regions within which β can be considered as a constant: *Jura* (β = 0.87); *Plateau* (β = 0.42); *Pre-Alps* (β = 0.87). This approach, as Niggli noted

(2004), is only a preliminary and not very satisfying attempt, and needs to be reworked to be fully operational. In particular, as the parameter β obtained is the only calibration parameter, it expresses the total explained variance and it reaches a value that has no plausible physical reality.

9.4.4 Final Comment

Following this quick overview of regional analysis, we can draw the following conclusions:

- in the case of runoff, the approach will often be different depending on the size of the watersheds being studied.
- it can often be indispensable to combine the approach using *explanatory variables*, the method using the *hypothesis of regional homogeneity*, and finally, the approach based on *spatial continuity* (for example, by using a kriging technique).

It should be added that some authors have proposed Bayesian methods of estimation in order to combine the estimations from these different approaches (Niggli, 2004; Niggli and Musy, 2003).

Bibliography

Abdous B., Genest C. and Rémillard B., Dependence properties of meta-elliptical distributions, *In*: P. Duchesne and B. Rémillard (editors), *Statistical modeling and analysis for complex data problems*, GERAD 25th Anniversary Series, Springer, New York, pp. 1-15, 2005.

Adamson P.T., Metcalfe Q.V. and Parmentier B., Bivariate extreme value distributions: An application of the Gibbs sampler to the analysis of floods, *Water Resources Research*, 35(9): 2825-2832, 1999.

Ahmad M.I., Sinclair C.D. and Spurr B.D., Assessment of flood frequency models using empirical distribution function statistics, *Water Resources Research*, 24(8): 1323-1328, 1988.

Akaike H., Information theory as an extension of the maximum likelihood principle, *In*: B.N. Petrov and F. Csaki (editors), *2nd International symposium on information theory*, Budapest, Akademiai Kiado, pp. 267-281, 1973.

Ali M., Mikhail N. and Haq M., A class of bivariate distributions including the bivariate logistic, *Journal of Multivariate Analysis*, 8(3): 405-412, 1978.

Anderson R.L., *Distribution of the Serial Correlation Coefficient, The Annals of Mathematical Statistics*, Vol. 13: 1-13, 1942.

Bagdonavicius V., Malov S. and Nikulin M., Characterizations and parametric regression estimation in Archimedean copulas, *Journal of Applied Statistical Science*, 8(1): 137-154, 1999.

Baker V.R., Paleoflood hydrology and extraordinary flood events. *Journal of Hydrology*, 96(1-4): 79-99, 1987.

Beard L.R., Probability estimates based on small normal distribution samples, *Journal of Geophysical Research*, 65(7): 2143-2148, 1960.

Bénichou P. and Le Breton O., Prise en compte de la topograhie pour la cartograhie des champs pluviométriques statistiques, *La Météorologie*, 7ème série, 19: 23-34, 1987.

Ben-Zvi A. and Azmon B., Joint use of L-moment diagram and goodness-of-fit test: a case study of diverse series, *Journal of Hydrology*, 198(1-4): 245-259, 1997.

Berger J.O., *Statistical Decision Theory and Bayesian Analysis,* Springer-Verlag, New York, 617 pp., 1985.

Berger J.O., *Bayesian analysis: A look at today and thoughts of tomorrow,* Technical Report 99-30, ISDS Discussion Paper, Duke University, 1999.

Bernardo J.M. and Smith A.F.M., *Bayesian Theory.* Wiley, New York, 573 pp., 1994.

Bernier J. and Veron R., Sur quelques difficultés rencontrées dans l'estimation d'un débit de crue de probabilité donnée, *Revue de Statistique Appliquée,* 12(1): 25-48, 1964.

Bernier J., *Sur les utilisations des L-Moments en hydrologie statistique,* Communication personnelle, 1993.

Bernier J., Parent E. and Boreux J.-J., *Statistique de l'Environnement. Traitement Bayésien de l'Incertitude.* Eds Tec&Doc (Lavoisier), Paris, 363 pp., 2000.

Bessis J., *La probabilité et l'évaluation des risques,* Masson, Paris, 132 pp., 1984.

Bobée B., The log-Pearson Type 3 distribution and its application to hydrology, *Water Resources Research,* 11(6): 681-689, 1975.

Bobée B. and Robitaille R., Correction of biais in the estimation of the coefficient of skewness, *Water Resources Research,* 33(12): 2769-2780, 1975.

Bobée B. and Degroseillers L., *Ajustement des distributions Pearson type 3, gamma, gamma généralisée et log-Pearson,* INRS-Eau, rapport de recherche 105, 63 pp., 1985.

Bobée B. and Ashkar F., *Sundry Averages Methods (SAM) for estimating parameters of the log-Pearson type 3 distribution,* INRS-Eau, rapport de recherche R-251, 28 pp., 1988.

Bobée B. and Ashkar F., *The Gamma family and derived distributions applied in hydrology,* Water Resources Publications, Littleton, 203 pp., 1991.

Bois P., *Contrôle des séries chronologiques par étude du cumul des résidus,* Colloque et séminaires ORSTOM, pp. 89-99, 1986.

Bouyé E., Durrleman V., Nikeghbali A., Riboulet G. and Roncalli T., *Copulas for finance - A reading guide and some applications,* rapport technique, Groupe de recherche opérationnelle, Crédit Lyonnais, Paris, 70 pp., 2000.

Bozdogan H., Model selection and Akaike's information criterion (AIC): The general theory and its analytical extentions, *Psychometrika,* 52(3): 345-370, 1987.

Breymann W., Dias A. and Embrechts P., Dependence structures for multivariate high-frequency data in finance, *Quantitative Finance*, 3(1): 1-14, 2003.

Brunet-Moret Y., Statistiques de rangs, *Cahiers ORSTOM, Série hydrologie*, 10(2): 133-151, 1973.

Brunet-Moret Y., Homogénéisation des précipitations, *Cahiers ORSTOM, Série hydrologie*, 16(3-4): 147-170, 1979.

Burnham K.P. and Anderson D.R., *Model selection and multimodel inference: A practical information-theoretic approach*, 2nd edition, Springer, New York, 488 pp., 2002.

Castellarin A., Vogel R.M. and Matalas N.C., Multivariate probabilistic regional envelopes of extreme floods, *Journal of Hydrology*, 336(3-4): 376-390, 2007.

Cave B.M. and Pearson K., Numerical illustrations of the variate difference correlation method, *Biometrika*, 10(2-3): 340-355, 1914.

CEA, *Statistique appliquée à l'exploitation des mesures*, Tome 1, Masson, 296 pp., Paris, 1978.

Chow V.T., The log-probability law and its engineering applications, *Proceedings of the American Society of Civil Engineers*, 80(536): 1-25, 1954.

Chow V.T. (editor), *Handbook of applied hydrology*, McGraw-Hill, New York, 1468 pp., 1964.

Clayton D.G., A model for association in bivariate life tables and its application in epidemiological studies of familial tendency in chronic disease incidence, *Biometrika*, 65(1): 141-151, 1978.

Coles S.G. and Tawn J.A., A Bayesian analysis of extreme rainfall data, *Applied Statistics*, 45(4): 463-478, 1996.

Coles S., *An introduction to statistical modeling of extreme values*, Springer, London, 224 pp., 2001.

Cook D. and Johnson M., A family of distributions for modeling non-elliptically symmetric multivariate data, *Journal of the Royal Statistical Society: Series B*, 43(2): 210-218, 1981.

CTGREF (Centre Technique du Génie Rural, des Eaux et Forêts), *Utilisation de quelques tests en hydrologie et calcul d'intervalles de confiance*, Informations techniques, Cahier 31, 2, Groupement d'Antony, Division hydrologie - hydraulique fluviale, 4 pp., 1978.

Cuadras C.M. and Augé J., A continuous general multivariate distribution and its properties, *Communications in Statistics - Theory and Methods*, 10(4): 339-353, 1981.

Cunderlik J.M. and Burn D.H., Non-stationary pooled flood frequency analysis, *Journal of Hydrology*, 276(1-4): 210-223, 2003.

Cunnane C., Unbiased plotting positions - A review, *Journal of Hydrology*, 37(3-4): 205-222, 1978.

Cunnane C., *Statistical distributions for flood frequency analysis*, World Meteorological Organization, Operational hydrology report 33, WMO 718, 115 pp., Geneva, 1989.

Danielsson J. and De Vries C.G., Tail index and quantile estimation with very high frequency data, *Journal of Empirical Finance*, 4: 241-257, 1997.

Danielsson J., de Haan L., Peng L. and de Vries C.G., Using a Bootstrap Method to Choose the Sample Fraction in Tail Index Estimation, *Journal of Multivariate Analysis* 76: 226-248, 2001.

Dalrymple T., *Flood frequency analysis*, U.S. Geological Survey, Water supply paper 1543 A, 80 pp., 1960.

Dauphiné A., *Risques et catastrophes*, Armand Colin, Paris, 288 pp., 2001.

Davison A.C. and Smith R.L., Models for exceedances over high thresholds (with discussion), *Journal of the Royal Statistical Society* 52: 393-442, 1990.

Davison A.C. and Hinkley D.V., *Bootstrap methods and their application*, Cambridge series in statistical and probabilistic mathematics, 1, Cambridge University Press, Cambridge, 592 pp., 1997.

Davison A.C. and Ramesh N.I., Local likelihood smoothing of sample extremes, *Journal of the Royal Statistical Society: Series B*, 62(1): 191-208, 2000.

De Michele C., Salvadori G., Canossi M., Petaccia A. and Rosso R., Bivariate statistical approach to check adequacy of dam spillway, *Journal of Hydrologic Engineering*, 10(1): 50-57, 2005.

De Montmollin F., O*livier* R. and Zwahlen F., Utilisation d'une grille d'altitudes digitalisées pour la cartographie d'éléments du bilan hydrique, *Journal of Hydrology*, 44(3-4): 191-209, 1979.

De Souza P.B., Consuegra D. and Musy A., CODEAU: A database package for the assessment and analysis of hydrometeorological data, *In*: A. Verwey, A.W. Minns, V. Babović and Č. Maksimović (editors), *Hydroinformatics '94, Proceeding of the 1st International conference on hydroinformatics*, Balkema, Rotterdam, pp. 103-110, 1994.

El Adlouni S., Ouarda T.B.M.J. and Bobée B., Orthogonal Projection L-moment estimators for three-parameter distributions, *Advances and Applications in Statistics*, sous presse, 2007.

Embrechts P., Klüppelberg C. and Mikosch T., *Modelling extremal events for insurance and finance*, Springer, Berlin, 655 pp., 1997.

Embrechts P., McNeil A. and Straumann D., Correlation and dependence in risk management: Properties and pitfalls, *In*: M. Dempster (editor), *Risk management: Value at risk and beyond*, Cambridge University Press, Cambridge, pp. 176-223, 2002.

Embrechts P., Lindskog F. and McNeil A., Modelling dependence with copulas and applications to risk management, *In*: S. Rachev (editor), *Handbook of heavy tailed distributions in finance*, Elsevier, New York, chapter 8, pp. 329-384, 2003.

Fang H.B., Fang K.T. and Kotz S., The meta-elliptical distributions with given marginals, *Journal of Multivariate Analysis*, 82(1): 1-16, 2002.

Favre A.C., Musy A. and Morgenthaler S., Two-site modeling of rainfall based on the Neyman-Scott process, *Water Resources Research*, 38(12): 1307, doi:10.1029/2002WR001343, 2002.

Favre A.C., El Adlouni S., Perreault L., Thiémonge N. and Bobée B., Multivariate hydrological frequency analysis using copulas, *Water Resources Research*, 40: W01101, doi:10.1029/2003WR002456, 2004.

Fermanian J.D., Goodness-of-fit tests for copulas, *Journal of Multivariate Analysis*, 95(1): 119-152, 2005.

Fortin V., Bernier J. and Bobée B., Rapport final du projet C3: *Détermination des crues de conception*, Chaire en hydrologie statistique, INRS-Eau, rapport R-532, 1998.

Frank M.J., On the simultaneous associativity of $F(x,y)$ and $x+y-F(x,y)$, *Aequationes Mathematicae*, 19(1): 194-226, 1979.

Frees E.W. and Valdez E.A., Understanding relationships using copulas, *North American Actuarial Journal*, 2(1): 1-25, 1998.

Galambos J., Order statistics of samples from multivariate distributions, *Journal of the American Statistical Association*, 70(3): 674-680, 1975.

Garrido M., *Modélisation des événements rares et estimation des quantiles extrêmes : méthodes de sélection de modèles pour les queues de distribution*, Thèse de doctorat, Université Joseph Fourier - Grenoble I, Grenoble, 231 pp., 2002.

Gelman A., Carlin J.B., Stern H.S and Rubin D.B., *Bayesian Data Analysis*, 2nd Edition, Chapman and Hall, Londres, 698 pp., 2004.

Genest C. and MacKay R., The joy of copulas: Bivariate distribution with uniform marginals, *The American Statistician*, 40: 280-283, 1986.

Genest C., Frank's family of bivariate distributions, *Biometrika*, 74(3): 549-555, 1987.

Genest C. and Rivest L.P., Statistical inference procedures for bivariate Archimedean copulas, *Journal of the American Statistical Association*, 88(423): 1034-1043, 1993.

Genest C. and Ghoudi K., Une famille de lois bidimensionnelles insolite, *Comptes Rendus de l'Académie des Sciences de Paris,* 318(1): 351-354, 1994.

Genest C., Ghoudi K. and Rivest L.P., A semiparametric estimation procedure of dependence parameters in multivariate families of distributions, *Biometrika,* 82(3): 543-552, 1995.

Genest C., Quessy J.F. and Rémillard B., Goodness-of-fit procedures for copula models based on the probability integral transformation, *Scandinavian Journal of Statistics,* 33(2): 337-366, 2006.

Genest C. and Favre A.C., Everything you always wanted to know about copula modeling but were afraid to ask, *Journal of Hydrologic Engineering,* 12(4): 347-368, 2007.

Genest C. and Rémillard B., Validity of the parametric bootstrap for goodness-of-fit testing in semiparametric models, *Annales de l'Institut Henri Poincaré,* 37, sous presse, 2007.

Genest C., Favre A.C., Béliveau J. and Jacques C., Meta-elliptical copulas and their use in frequency analysis of multivariate hydrological data, *Water Resources Research,* 43, W09401, doi:10.1029/2006WR005275, 2007.

Gibbons J.D. and Chakraborti S., *Nonparametric statistical inference,* 4th edition, Statistics: A Dekker series of textbooks and monographs, Marcel Dekker, New York, 680 pp., 2003.

Grandjean P., *Choix et élaboration des données de pluie pour l'assainissement rural et urbain dans le canton de Genève,* VSA, Journée « Niederschlag und Siedlungsentwäasserung », Zürich, 1988.

Greenwood J.A., Landwehr J.M., Matalas N.C. and Wallis J.R., Definition and relation to parameters of several distributions expressable in inverse form, *Water Resources Research,* 15(5): 1049-1054, 1979.

GREHYS (Groupe de Recherche en Hydrologie Statistique), Presentation andreview of some methods for regional flood frequency analysis, *Journal of Hydrology,* 186(1-4): 63-84, 1996.

Guillot P. and Duband D., La méthode du GRADEX pour le calcul de la probabilité des crues rares à partir des pluies, *In*: IAHS (editor), *Les crues et leur évaluation,* International Association of Hydrological Sciences, 84, pp. 560-569, 1967.

Guillot P., Les arguments de la méthode du gradex, base logique pour évaluer les crues extrêmes, *Symposium IDNDR,* Oklahoma, 1994.

Gumbel E.J., *Statistics of extremes,* Columbia University Press, New York, 375 pp., 1958.

Gumbel E.J., Bivariate exponential distributions, *Journal of the American Statistical Association,* 55(292): 698-707, 1960.

Gunasekara T.A.G. and Cunnane C., Expected probabilities of exceedance for non-normal flood distributions, *Journal of Hydrology*, 128(1-4): 101-113, 1991.

Harr M.E., *Lectures on reliability in civil engineering*, Swiss Federal Institute of Technology, Lausanne, 1986.

Harr M.E., Probabilistic estimates for multivariate analysis, *Appied Mathematical Modelling*, 13(5): 313-318, 1989.

Hiez G., L'homogénéité des données pluviométriques, *Cahiers ORSTOM, Série hydrologie*, 14(2): 129-172, 1977.

Hill, B.M., A simple general approach to inference about the tail of a distribution, *The Annals of Statistics*, 3(5): 1163-1174, 1975.

Hosking J.R.M., L-moments: Analysis and estimation of distributions using linear combinations of order statistics, *Journal of the Royal Statistical Society: Series B*, 52(1): 105-124, 1990.

Hosking J.R.M. and Wallis J.R., Some statistics useful in regional frequency analysis, *Water Resources Research*, 29(2): 271-281, 1993.

Hörler A., *Die Intensitäten von Starkregen längerer Dauer für verschiedene ortschaften der Schweiz*, Gaz-Wasser-Abwasser, 57. Jahrgang, 12: 853-860, 1977.

Hougaard P., A class of multivariate failure time distributions, *Biometrika*, 73(3): 671-678, 1986.

Hüsler J. and Reiss R.D., Maxima of normal random vectors: between independence, and complete dependence, *Statistics and Probability Letters*, 7(4): 283-286, 1989.

Iman R.L. and Conover W.J., *A modern Approach to Statistics*, Wiley, New-York, 1983.

Joe H., Parametric families of multivariate distributions with given margins, *Journal of Multivariate Analysis*, 46(2): 262-282, 1993.

Joe H., *Multivariate models and dependence concepts*, Chapman and Hall, London, 399 pp., 1997.

Jordan J.P. and Meylan P., *Estimation spatiale des précipitations dans l'ouest de la Suisse par la méthode du krigeage*, IAS, 12: 157-162, 1986a.

Jordan J.P. and Meylan P., *Estimation spatiale des précipitations dans l'ouest de la Suisse par la méthode du krigeage*, IAS, 13: 187-189, 1986b.

Katz R.W., Parlange M.B. and Naveau P., Statistics of extremes in hydrology, *Advances in Water Resources*, 25(8-12): 1287-1304, 2002.

Kendall M.G., A new measure of rank correlation, *Biometrika*, 30(1-2): 81-93, 1938.

Kendall M.G., Stuart A., Ord K. and Arnold S., *Kendall's advanced theory of statistics*, Volume 2A: Classical inference and the linear models, 6th edition, Arnold, London, 448 pp., 1999.

Khaliq M.N., Ouarda T.B.M.J., Gachon P. and Bobée B., Frequency analysis of a sequence of dependent and/or non-stationary hydro-meteorological observations: A review, *Journal of Hydrology*, 329(3-4): 534-552, 2006.

Kimeldorf G. and Sampson A.R., Uniform representations of bivariate distributions. *Communications in Statistics: Theory and Methods*, 4(7): 617-627, 1975.

Kite G.W., Confidence limits for design event, *Water Resources Research*, 11(1): 48-53, 1975.

Kite G.W., *Frequency and risk analysis in hydrology*, Water Resources Publications, Littelton, 224 pp., 1988.

Klemeš V., Dilletantism in hydrology: Transition or destinity?, *Water Resources Research*, 22(9): 177S-188S, 1986.

Klemeš V., Tall tales about tails of hydrological distributions. II, *Journal of Hydrologic Engineering*, 5(3): 232-239, 2000.

Koenker R. and Basset G., Regression quantiles, *Econometrica*, 46(1): 33-50, 1978.

Koenker R. and D'Orey V., Computing regression quantiles, *Applied Statistics*, 36(3): 383-393, 1987.

Koutsoyiannis D., Kozonis D. and Manetas A., a mathematical framework for studying rainfall intensity-duration-frequency relationships, *Journal of Hydrology*, 206: 118-135, 1998.

Kuczera G., Combining site-specific and regional information: an empirical Bayes approach, *Water Resources Research*, 18(2): 306-314, 1982.

Kuczera G., Comprehensive at-site flood frequency analysis using Monte Carlo Bayesian inference, *Water Resources Research*, 34(6): 1551-1558, 1999.

Landwehr J.M. and Matalas N.C., Probability weighted moments compared with some traditional techniques in estimating Gumbel parameters and quantiles, *Water Resources Research*, 15(5): 1055-1064, 1979.

Lang M., *Communication personnelle*, CEMAGREF, Groupement de Lyon, 1996.

Langbein W.B., Annual floods and the partial-duration flood series, *Transactions of the American Geophysical Union*, 30(6): 879-881, 1949.

Larras J., *Prévision et prédétermination des crues et des étiages*, Eyrolles, Paris, 160 pp., 1972.

Lebart L., Morineau A. and Fenelon J.P., *Traitement des données statistiques*, 2ème édition, Dunod, Paris, 510 pp., 1982.

Lebel T. and Boyer J.F., *DIXLOI : Un ensemble de programmes Fortran 77 pour l'ajustement de lois statistiques et leur représentation graphique*, ORSTOM, Laboratoire d'hydrologie, Montpellier, 55 pp., 1989.

Linsley R.K., Flood estimates: How good are they?, *Water Resources Research*, 22(9): 159S-164S, 1986.

Mallows C.L., Some comments on C_p, *Technometrics*, 15(4): 661-675, 1973.

Masson J.M., *Méthode générale approchée pour calculer l'intervalle de confiance d'un quantile. Application à quelques lois de probabilité utilisées en hydrologie*, Note interne 6/1983, Laboratoire d'hydrologie et modélisation, Université Montpellier II, 41 pp., 1983.

Matheron G., *Estimer et choisir. Essai sur la pratique des probabilités*, Cahiers du Centre de morphologie mathématique de Fontainebleau, 7, Ecole Nationale Supérieure des Mines de Paris, ENSMP, Fontainebleau, 175 pp., 1978.

McNeil A.J. and Saladin T., Developing scenarios for future extreme losses using the POT method, *In*: P. Embrechts (editor), *Extremes and integrated risk management*, Risk Books, London, 397 pp., 2000.

Meylan P., Régionalisation de données entachées d'erreurs de mesure par krigeage, *Hydrologie Continentale*, 1(1): 25-34, 1986.

Meylan P. and Musy A., *Hydrologie fréquentielle*, HGA, Bucarest, 413 pp., 1999.

Meylan P., Grandjean P. and Thöni M., *Intensité des pluies de la région genevoise – Directive IDF 2001*, Etat de Genève, DIAE, DomEau, SECOE, 4 pp., 2005.

Meylan P., *A.F.P.C. – Analyse fréquentielle par comptage*, République et canton de Genève – Département de l'intérieur et de la mobilité – Direction générale de l'eau, AIC, Lausanne, février 2010.

Michel C., *Hydrologie appliquée aux petits bassins ruraux*, CEMAGREF, Division hydrologie, hydraulique fluviale et souterraine, Antony, 528 pp., 1989.

Miquel J., *Guide pratique d'estimation des probabilités de crues*, Eyrolles, Paris, 160 pp., 1984.

Morel-Seytoux H.J., Forecasting of flow-flood frequency analysis, *In*: H.W. Shen (editor), *Modeling of rivers*, chapter 3, Wiley, New York, 1979.

Morgenthaler S., *Introduction à la statistique*, Méthodes mathématiques pour l'ingénieur, 9, 2ème édition, Presses polytechniques et universitaires romandes, Lausanne, 328 pp., 2001.

Morlat G., Les lois de probabilités de Halphen, *Revue de statistique appliquée,* 4(3): 21-43, 1956.

Musy A. and Higy C., *Hydrology: A Science of Nature,* Science Publishers, Enfield (USA), 326 pp., 2010.

Naulet R., *Utilisation de l'information des crues historiques pour une meilleure prédétermination du risque d'inondation. Application au bassin de l'Ardèche à Vallon Pont-d'Arc et Saint Martind' Ardèche.* PhD thesis, Université Joseph Fourier - Grenoble 1 (France) et INRS-Eau Terre et Environnement (Québec), 2002.

Nelsen R.B., Properties of a one-parameter family of bivariate distributions with specified marginals, *Communications in Statistics: Theory and Methods,* 15(11): 3277-3285, 1986.

Nelsen R.B., *An introduction to copulas,* Lecture notes in statistics, 139, 2nd edition, Springer, New York, 269 pp., 2006.

NERC (Natural Environment Research Council), *Flood studies report,* Volume I: Hydrological studies, NERC, London, 1975.

NERC (Natural Environment Research Council), *Review of regional growth curves,* Flood studies supplementary report 14, London, 6 pp., 1983.

Neykov N.M., Neytchev P.N., Van Gelder P.H.A.J.M. and Todorov V.K., Robust detection of discordant sites in regional frequency analysis, *Water Resources Research,* 43: W06417, doi:10.1029/2006WR005322, 2007.

Niggli M. and Musy A., A Bayesian combination method of different models for flood regionalisation, *In: Proceedings of the 6th Inter-regional conference on environment-water, Land and water use planning and management,* Albacete, 2003.

Niggli M., *Combinaison bayésienne des estimations régionales des crues : concept, développement et validation,* Thèse de doctorat 2895, EPFL, Lausanne, 229 pp., 2004.

Oakes D., A model for association in bivariate survival data, *Journal of the Royal Statistical Society: Series B,* 44(3): 414-422, 1982.

Oakes D., Multivariate survival distributions, *Journal of Nonparametric Statistics,* 3(3-4): 343-354, 1994.

O'Connell D.R.H., Osteena D.A., Levish D.R. and Klinger R.E., Bayesian flood frequency analysis with paleohydrologic bound data. *Water Resources Research,* 38(5), 1058, doi:10.1029/2000WR000028, 2002.

OFEFP (Office fédéral de l'environnement, des forêts et du paysage), *Manuel I de l'ordonnance sur les accidents majeurs OPAM,* OFEFP, Berne, 74 pp., 1991a.

OFEFP (Office fédéral de l'environnement, des forêts et du paysage), *Débits de crue dans les cours d'eau suisses*, Communications hydrologiques, 3 et 4, OFEFP, Berne, 1991b.

OFEG (Office fédéral des eaux et de la géologie), *Protection contre les crues des cours d'eau – Directives 2001*, Bienne, OFEG, 72 pp., 2001.

OFEG (Office fédéral des eaux et de la géologie), *Evaluation des crues dans les bassins versants de Suisse - Guide pratique*, Rapports de l'OFEG, Série Eaux, 4, OFEG, Berne, 114 pp., 2003.

Ouarda T., Rasmussen P.F., Bobée B. and Bernier J., Utilisation de l'information historique en analyse hydrologique fréquentielle, *Revue des sciences de l'eau*, 11(1): 41-49, 1998.

Ouarda T.B.M.J., Cunderlik J., St-Hilaire A., Barbet M., Bruneau P. and Bobée B., Data-based comparison of seasonality-based regional flood frequency methods, *Journal of Hydrology*, 330(1-2): 329-339, 2006.

Parent E. and Bernier J., Encoding prior experts judgments to improve risk analysis of extreme hydrological events via POT modeling. *Journal of Hydrology*, 283(1-4): 1-18, 2003.

Parent E. and Bernier J., Bayesian POT modeling for historical data. *Journal of Hydrology*, 274(1-4): 95-108, 2003.

Parent E. and Bernier J., *Le raisonnement bayésien. Modélisation et inférence.* Springer, Statistique et probabilités appliquées, Paris, 364 pp., 2007.

Payrastre O., *Faisabilité et utilité du recueil de données historiques pour l'étude des crues extrêmes de petits cours d'eau. Etude du cas de quatre bassins versants affluents de l'Aude*. Thèse. École nationale des ponts et chaussées (France), 2005.

Plackett R.L., A class of bivariate distributions, *Journal of the American Statistical Association*, 60(2): 516-522, 1965.

Peck E.L. and Brown M.J., An approach to the development of isohyetal maps for montainous areas, *Journal of Geophysical Research*, 67(2): 681-694, 1962.

Perreault L., *Analyse bayésienne rétrospective d'une rupture dans les séquences de variables aléatoires hydrologiques*. Thèse de doctorat en statistique appliquée, ENGREF, Paris, 2000.

Perreault L., *Modélisation des débits sortants à Chute-des-Passes : application du mélange de distributions normales*, Rapport Institut de Recherche d'Hydro-Québec. IREQ-2002-127C, confidentiel, 41 pp, 2003.

Pickands III J., Statistical Inference Using Extreme Order Statistics, *The Annals of Statistics*, 3(1): 119-131, 1975.

Poulin A., Huard D., Favre A.C. and Pugin S., Importance of tail dependence in bivariate frequency analysis, *Journal of Hydrologic Engineering*, 12(4): 394-403, 2007.

Proschan F., Confidence and tolerance intervals for the normal distribution, *Journal of the American Statistical Association*, 48(263): 550-564, 1953.

Pugachev V.S., *Théorie des probabilités et statistique mathématique*, MIR, Moscou, 471 pp., 1982.

Ramesh N.I. and Davison A.C., Local models for exploratory analysis of hydrological extremes, *Journal of Hydrology*, 256(1-2): 106-119, 2002.

Rao D.V., Estimating log Pearson parameters by mixed moments, *Journal of Hydraulic Engineering*, 109(8): 1118-1132, 1983.

Reiss R.D. and Thomas M., *Statistical analysis of extreme values*, 2nd edition, Birkhäuser, Bâle, 443 pp., 2001.

Restle E.M., El Adlouni S., Bobée B. and Ouarda T.B.M.J., *Le test GPD et son implémentation dans le logiciel HYFRAN PRO*, INRS-ETE, rapport de recherche R-737, 35 pp., 2004.

Ribatet M., Sauquet E., Grésillon J.M. and Ouarda T.B.M.J., A regional Bayesian POT model for flood frequency analysis, *Stochastic Environmental Research and Risk Assessment*, 21(4): 327-339, 2007.

Rissanen J., Modeling by shortest data description, *Automatica*, 14(5): 465-471, 1978.

Robert C.P. and Casella G., *Monte Carlo Statistical Methods*, 2nd Edition, Springer-Verlag, New York, 635 pp., 2004.

Rosenblueth E., Point estimates for probability moments, *Proceedings of the National Academy Sciences of the United States of America*, 72(10): 3812-3814, 1975.

Ross S.-M., *An Introduction to Probability Models*, 10th Edition, Academic Press, Elsevier, Burlington, USA, 784 pp., 2010.

Sachs L., *Applied statistics: A handbook of techniques*, 2nd edition, Springer, New York, 706 pp., 1984.

Salvadori G. and De Michele C., Frequency analysis via copulas: Theoretical aspects and applications to hydrological events, *Water Resources Research*, 40: W12511, doi:10.1029/2004WR003133, 2004.

Sankarasubramanian A. and Lall U., Flood quantiles in a changing climate: Seasonal forecasts and causal relations, *Water Resources Research*, 39(5): 1134, doi:10.1029/2002WR001593, 2003.

Saporta G., *Probabilités, analyse des données et statistique*, Technip, Paris, 493 pp., 1990.

Schwarz G., Estimating the dimension of a model, *The Annals of Statistics*, 6(2): 461-464, 1978.

Schweizer B. and Wolff E., On nonparametric measures of dependence for random variables, *The Annals of Statistics*, 9(4): 879-885, 1981.

Sherrod Ph.-H., *NLREG – Nonlinear regression analysis Program*, Brentwood, 2002.

Shih J.H. and Louis T.A., Inferences on the association parameter in copula models for bivariate survival data, *Biometrics*, 51(4): 1384-1399, 1995.

Siegel S., *Nonparametric statistics for the behavioral sciences*, 2nd edition, McGraw-Hill, New York, 399 pp., 1988.

Sklar A., Fonctions de répartition à *n* dimensions et leurs marges, *Publications de l'Institut de statistique de l'Université de Paris*, 8: 229-231, 1959.

Sneyers R., *Sur l'analyse statistique des séries d'observations*, Note technique 143, OMM, Genève, 192 pp., 1975.

Sokolov A.A., Rantz S.E. and Roche M.F., *Floodflow computation: Methods compiled from world experience*, Studies and reports in hydrology, 22, UNESCO Press, Paris, 294 pp., 1976.

Solaiman B., *Processus stochastiques pour l'ingénieur*, Collection technique et scientifique des télécommunications, Presses polytechniques et universitaires romandes, Lausanne, 241 pp., 2006.

Stedinger J.R., Vogel R.M. and Foufoula-Georgiou E., Frequency analysis of extreme events, chapter 18, *In*: D.R. Maidment (editor), *Handbook of Hydrology*, McGraw-Hill, New York, 1424 pp., 1993.

Stephens M.A., EDF statistics for goodness of fit and some comparisons, *Journal of the American Statistical Association*, 69(347): 730-737, 1974.

Strupczewski W.G. and Kaczmarek Z., Non-stationary approach to at-site flood frequency modelling II. Weighted least squares estimation, *Journal of Hydrology*, 248(1-4): 143-151, 2001.

Strupczewski W.G., Singh V.P. and Feluch W., Non-stationary approach to at-site flood frequency modelling I. Maximum likelihood estimation, *Journal of Hydrology*, 248(1-4): 123-142, 2001a.

Strupczewski W.G., Singh V.P. and Mitosek H.T., Non-stationary approach to at-site flood frequency modelling III. Flood analysis of Polish rivers, *Journal of Hydrology*, 248(1-4): 152-167, 2001b.

Sturges H., The choice of a class-interval. *Journal of the American Statistical Association*, 21: 65-66, 1926.

Takeuchi K., Annual maximun series and partial-duration series — Evaluation of Langbein's formula and Chow's discussion, *Journal of Hydrology*, 68(1-4): 275-284, 1984.

Tawn J.A., Bivariate extreme value theory: Models and estimation, *Biometrika*, 75(3): 397-415, 1988.

Thirriot C., Quoi de neuf depuis Noé ? *In*: *23èmes Journées de l'hydraulique*, Rapport général, Société hydrotechnique de France, Nîmes, 1994.

Ventsel H., *Théorie des probabilités*, MIR, Moscou, 563 pp., 1973.

Viglione A., Laio F. and Claps P., A comparison of homogeneity tests for regional frequency analysis, *Water Resources Research*, 43: W03428, doi:10.1029/2006WR005095, 2007.

Vogel R.M. and Fennessey N.M., L moment diagrams should replace product moment diagrams, *Water Resources Research*, 29(6): 1745-1752, 1993.

VSS (Association suisse des professionnels de la route et des transports), *Evacuation des eaux de chaussée - Intensité des pluies*, Norme SN 640 350, Zürich, 8 pp., 2000.

Wald A. and Wolfowitz J., An Exact Test for Randomnesss in the Non-Parametric Case based on serial Correlation, *The Annals of Mathematical Statistics*, Vol. 14: 378-388, 1943.

Wallis J.R., Catastrophes, computing, and containment: Living in our restless habitat, *Speculation in Science and Technology*, 11(4): 295-315, 1988.

Weiss L.L., Ratio of true to fixed interval maximum rainfall, *Journal of the Hydraulics Division ASCE*, 90(1): 77-82, 1964.

WMO (World Meteorological Organisation), Guide to hydrological practices Volume II : Analysis, forecasting and other applications, 168, OMM, Geneva, 304 pp., 1981.

WRC (U.S. Water Resources Council), *Guidelines for determining flood flow frequency*, Hydrology committee, WRC, Washington, 1967.

Yevjevitch V., *Probability and statistics in hydrology*, Water Resources Publications, Fort Collins, 302 pp., 1972.

Yue S., Ouarda T.B.M.J. and Bobée B., A review of bivariate gamma distribution for hydrological application, *Journal of Hydrology*, 246(1-4): 1-18, 2001.

Yue S. and Rasmussen P., Bivariate frequency analysis: Discussion of some useful concepts in hydrological application, *Hydrological Processes*, 16(14): 2881-2989, 2002.

Zhang X., Zwiers F.W. and Li G., Monte Carlo experiments on the detection of trends in extreme values, *Journal of Climate*, 17(10): 1945-1952, 2004.

List of Acronyms

AIC 116, 117	Akaike Information Criterion
AIC ix	AIC Ingénieurs conseils S.A.
ANETZ 32	Swiss Network of Automatic Weather Stations
BIC 116, 117	Bayesian Information Criterion
CEA 156	Commissariat à l'Énergie Atomique
CEMAGREF 111	Centre de Machinisme Agricole, Génie Rural Eaux et Forêts
CODEAU 37	CODification Eau (software package)
CTGREF 111, 112, 125	Centre Technique du Génie Rural, Eaux et Forêts
ECOL ix	Ecological Engineering Laboratory
ENSO 178	El-Nino-Southern Oscillation
EPFL vi, ix	Ecole Polytechique Fédérale de Lausanne
EXP 79, 143	EXPonential distribution
GAM 80	GAMma distribution
GEV **76**, 79, 85, 114, **143**, 177	Generalized Extreme Value distribution
GPD 79, **143**	Generalized Pareto Distribution
GUM **1**, 79, 143	GUMbel distribution
HYDRAM viii	HYDrologie et AMénagements (Hydrology and Land Improvement laboratory)
IDF 135, 145, 146, 149, 185	Intensity - Duration - Frequency curve
L 86	L-moment
LN 80	Lognormal distribution
LN3 85, 86, 88, 89	3-parameter LogNormal distribution
MCMC 164	Markov Chain Monte Carlo method
MDL 116	Minimum Description Length

Index and List of Proper Names

Subject Index

9781578087471